PARADOX NULL

The History of Paradoxes and the Laws that End Paradoxes

James O Harris

CONTENTS

INTRODUCTION

There are paradoxes in physics today. Far more paradoxes than most scientists will ever realize. Five classes of paradoxes prevail: paradoxes of singularities and black holes; paradoxes of stellar fusion; paradoxes of light; paradoxes of gravity (dark energy and dark matter); and paradoxes of big bang and expansion theories.

All of these theories have one thing in common: Astrophysics. All of these physics have one descriptor in common: Theoretical. Since most of those physics happen not around here, they are not commonly disputed.

All five are key problems in the quest for a universal set of physics.

The singularity paradoxes include this problem: Mass and gravity do not exist without matter (atomic nuclei). Singularities are supposed to be smaller than atoms, so to get a singularity, matter must be destroyed. But when matter is destroyed, there is no mass or gravity. That leaves no force to form any singularity and no force to hold that nothing together.

Black holes involve many paradoxes including gravity paradoxes.

The first paradox of black holes is the idea of black holes stopping time. Gravity is a function of time. Light speed is also a function of time. If black holes stopped time, the region in which time has stopped cannot exert gravity. If black holes stopped time, nothing could happen there - no rotation, no emissions, no interactions, and no collisions.

All of those things happen anyhow.

Stellar fusion paradoxes include the Massachussetts Institute of Technology (MIT) hot fusion project which has replicated conditions in which fusion is thought occur. It doesn't. MIT's project is a continual failure.

Stellar fusion paradoxes involve the solar wind. The solar wind consists of hydrogen, helium, and other gases which evaporate from the Sun and form a wind blowing past all of the planets. That's a huge problem because if those gases can evaporate from the Sun, then the pressures on the Sun are exponentially too low to cause fusion.

Then there are the fusion chain paradoxes. First of all, atoms tend to fission, not fuse, when exposed to extreme conditions. The foundational concept of hot fusion is erroneous. A key fusion paradox is that heavier elements including iron cannot be formed in hot fusion.

The paradoxes of light include 'wave-particle duality'. The principle of wave-particle duality is a scientific shrug at the paradox that light cannot, in a persistent form, behave as sometimes a wave and sometimes a particle. The concept of wave-particle duality is that this does happen, that this is impossible, and they don't know why.

The paradoxes of gravity include dark energy and dark matter - two substances which do not exist according to three large independent experiments which occurred in October and November of 2016. Classic theory observes that the universe does not obey gravity as defined by Isaac Newton. No individual galaxy obeys gravity as defined by Isaac Newton. Therefore, Newton's gravity is either incorrect or incomplete.

Dark energy and dark matter are theories which attempt(ed) to complete Newton's Law of Gravity. This book demonstrates that the true solution to both of the 'dark' gravity problems is the Law of Time.

The paradoxes of expansion theory are connected with with the gravity paradoxes and include dark energy as a theoretical driving force.

Here are the fundamental, erroneous assumptions of modern science: Space is nothing; light is flying photons, mass takes up no space; stars are made of hydrogen; stars are powered by fusion; and the whole universe shares a single, homogeneous inertial frame of reference.

Part I of Paradox Null is a history of physics and paradoxes including how everything basically works.

Part II of Paradox Null shows how practical physics make many strange theories impracticle. two new laws, shows how these combine to form a universal set of physics.

Part II of Paradox Null demonstrates two new laws of physics and shows how those principles eliminate scores of paradoxes.

There is only one fundamental rule in the Paradox Null set of Universal Physics:

When physics obey physics, there are no paradoxes.

THE PARADOX

A man approaches you and says, "Everything I say is a lie."

That statement cannot be true.

If that statement were true, then the statement, "Everything I say is a lie" must be false. For if a man always did lie, then he would have to lie to you about always lying.

But he said, "Everything I say is a lie." That cannot be true.

If everything that man says is not a lie, then the statement "Everything I say is a lie" is a lie.

Therefore, the statement "Everything I say is a lie" must be false.

This statement proves itself false. It is a logical paradox. It all adds up to 'that guy is a liar and everything he says is not a lie.'

Paradoxes contradict themselves.

Paradoxes are always false.

Dedicated to my wonderful wife, Rawlene LeBaron

"I'm just getting started."

PART I: THE HISTORY OF PHYSICS AND PARADOXES

In this section, the history of physics is summarized along with the people, the places, and the paradoxes that emerged. Part I also contains new physical principles concerning wave/particle relationships.

If the reader does not have interest in science history, science people, or context of historical paradoxes, Part II: Paradox Null is focused primarily on new physics and new physics laws including a new definition of space-time and demonstrations of laws that resolve many paradoxes.

Physics Beyond the Stone Age

We might think of the modern world of science as beginning around 300 BC – about 2,300 years ago. That is when Euclid (323–283 BC) reasoned, deduced, and visually, verbally, and numerically expressed *Euclidean Math* in a text titled *The Elements*. Euclidean math includes the equations for working with squares, circles, and triangles.

Euclidean Math established everything needed for levers, pulleys, and the building of Greece and Rome.

The Elements establishes basic algebra and geometry. These concepts lay the foundations for trigonometry and calculus (*non-euclidean math*).

In 1235 French architect Villiard De Honnecourt designed the overbalanced wheel. This intriguing device generates rotational motion from gravity. The overbalanced wheel pictured in this picture rotates clockwise.

This works because gravity accelerates all the rollers downwards at the same rate. They all weigh the same, but the

Overbalanced Wheel
working example
https://www.youtube.com/watch?v=287qd4ul7-F

roller on the right side of the wheel is always farther away from the center than the roller on the left. Therefore, the center of gravity is always on the right side of the center. Because of this, the right side always falls, and this overbalanced wheel spins clockwise.

Leonardo da Vinci

Leonardo di ser Piero da Vinci (1452 – 1519) was an Italian polydisciplinary genius whose interests included natural philosophy, invention, engineering, mathematics, drawing, painting, sculpting, architecture, music, literature, anatomy, geology, cartography, astronomy, botany, language, and history. Da Vinci is the definition of *Renaissance Man* and he was a pivotal figure of the *Renaissance Era*.

Da Vinci was one of 12 children and was home-schooled in Latin, geometry, and math. At age 14, da Vinci was taken as an apprentice by artist Andrea di Cione (1435 - 1488), known as Verrocchio. At the age of 20, Leonardo became a Master in the Guild of Saint Luke – a guild of artists and doctors of medicine.

Leonardo worked extensively with the overbalanced wheel. Due to Leonardo's many drawings of it, it is sometimes called the da Vinci wheel.

Da Vinci's studies, coupled with the fantastic ability to visually express concepts, made da Vinci's works essential to the oncoming world of *science*. Branches of what we now call science which da Vinci might be credited with originating include paleontology, ichnology (microfossils and indirect fossils), and architecture.

Nicolaus Copernicus (1473 – 1543) was a Renaissance-era Polish mathematician and astronomer famous for rearranging the astronomical world. Before Copernicus, the prevailing model of the cosmos is that everything orbited the Earth.

From the surface of the Earth, all objects in the sky appear to orbit the Earth, so Earth can logically be interpreted as being the central point. The concept of the universe orbiting the Earth is known as *geocentrism.*

In 1543, thirty years after he had originated a different cosmic model and just before his death, Copernicus published *On the Revolutions of the Celestial Spheres* which described his new cosmic model and listed seven key assumptions of his theory, including:

2. The center of the earth is not the center of the universe, but only the center towards which heavy bodies move and the center of the lunar sphere.
3. All the spheres surround the sun as if it were in the middle of them all, and therefore the center of the universe is near the sun.

This concept is known as heliocentrism. Helios meaning 'Sun', heliocentrism is the belief that the Sun is (near) the center of the universe.

The Copernicus Paradox

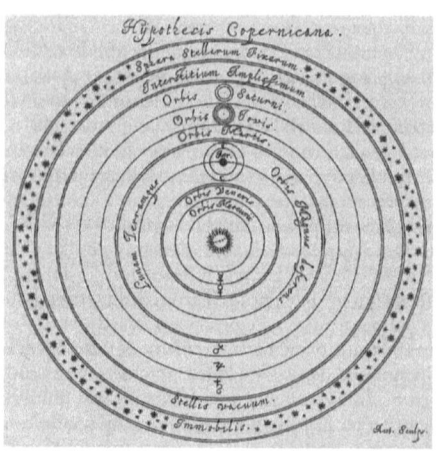

One of the problems that Copernicus faced was that all of his planets existed in spheres. For that to be true, all orbits must be circular. We now know that none of the planets in the Solar System have perfectly circular orbits.

Assuming them to be circular, Copernicus had a hard time figuring out what the common center would be. There is no common center. There is no solution to this problem.

According to the Copernicus model of the Solar System, the Earth, for instance, does not orbit the Sun, but instead orbits some point of empty space 2.5 billion kilometers away from the Sun as graphed below where if Earth's orbit were circular, the actual point in space that Earth would be orbiting is pretty far away from the Sun. The Sun is magnified fifty times in this image.

The Copernicus Paradox

Center of Earth's orbit ↓

Sun magnified 50x ●

If Earth's orbit is circular, Earth does not orbit the Sun.

3,593 Solar Radii
(aphelion - perihelion) /2 = 2,502,500,000 kilometers

This is the Copernicus Paradox. Using the Copernicus model, the Earth does not orbit the Sun and none of the other planets do, either. Every one orbits a different place somewhere near the Sun. This is why Copernicus said that the center of the universe was 'near the Sun' instead of identifying the Sun directly.

One aspect of Copernicus' model has prevailed: 'The center of the Earth... the center towards which heavy bodies move...' - this concept was later used by Isaac Newton and is still in use today. *Copernicun Gravity* is also used as the standard model of atomic structures.

The world of science now hinged on the assumption that objects are gravitationally attracted to other objects from the centers of those objects.

In the year 1600 William Gilbert (1544 – 1603) published *On the Magnet and Magnetic Bodies, and on the Great Magnet the Earth.* Making Gilbert a important contributor to the theories of magnetism and the new field of *electricus* – he being one of the first men to use that term. Electricus was a New Latin term meaning *like amber,* referencing the spark discharges he and colleagues were working with..

In 1646 electricus was supplanted with the term *electricity*

Gilbert submitted that electricity and magnetism were not the same thing, pointing out that electricity dissipates but magnetism does not. Electrical transmission is greatly affected by temperature, but magnetism is not.

Refraction of Light

In 1621 Oubichapter Snellius (1580 – 1626) determined the *law of refraction* or *Snell's Law.* Snellius – or simply Snell in English – was a Dutch astronomer, mathematician, and surveyor.

Snell was born in Leiden, Netherlands. Snell's father, Rudolph Snel van Royen (1546–1613) was professor of mathematics at the University of Leiden – a position Snell later filled.

Snell determined that the angle of refraction of light entering a medium is proportional to the refraction angle of light when it leaves the medium. Snell's law applies to refractive media such as glass and mathematically expresses how water, lenses, and prisms bend light. Water bends light because, light travels 32.8% faster in air than in water, and 17.6% slower in glass than in water.

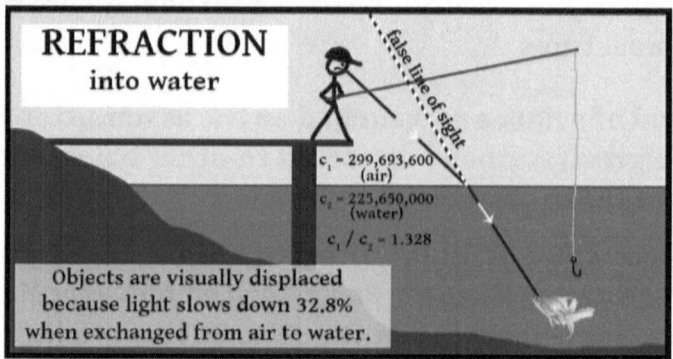

This picture demonstrates how the lensing of light by water causes a man viewed by a fish to appear in a different place than it actually is. The fish is also visually displaced, as indicated by the man's hook.

As seen in the the *Refraction illustration* below, the waves of light from water spread when they are transferred to the air. This happens because they travel faster in air than in water. This changing of the speed of light lenses the natural path of light. The path of light changes when crossing from one medium to another, forming two different paths.

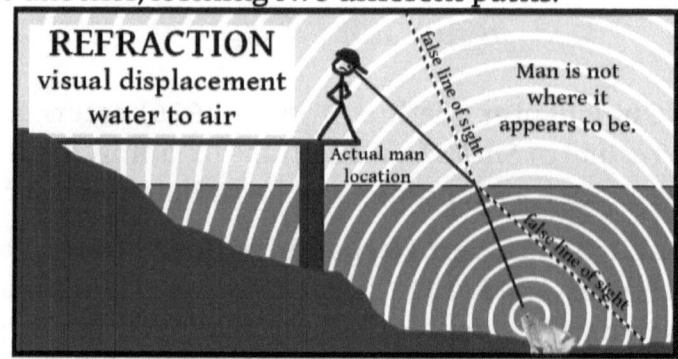

As shown in *Refraction, same angle different displacement*, light-waves spread when transferred from water to air. Because of this dilation, the man will appear farther away than he actually is, but from the man's perspective, the fish appears to be closer than it actually is.

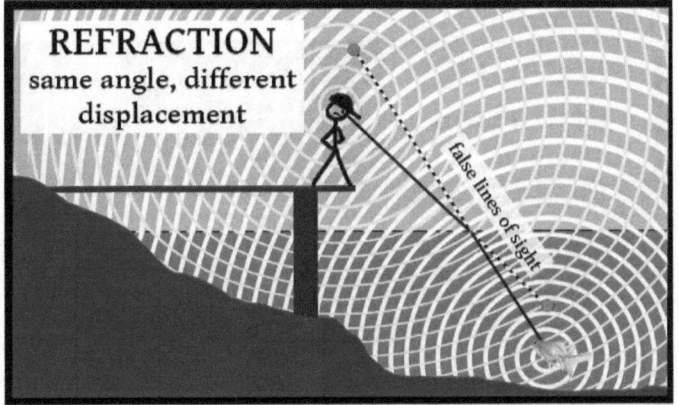

The illustration below shows Snell's law applied to the fish conundrum. The refraction indexes – n_1 and n_2 – are the proportional speeds of light in the differing mediums.

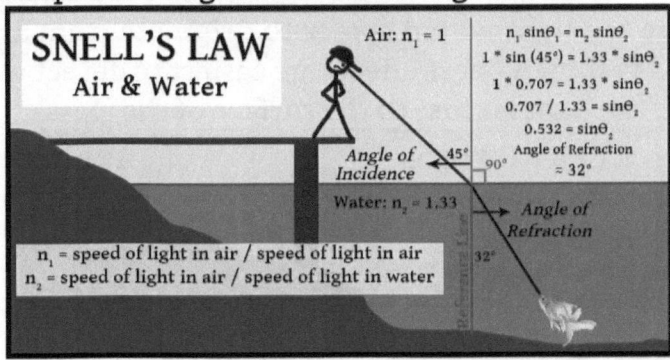

The path of light is bent unless that path is perpendicular to the surface. In that case – if the fish were directly below the fisherman – the path of light would be straight and the fish would be seen at an accurate vertical location, although it will still be deeper than its appearance indicates. The vertical path of light through a refractive surface is called the *reference line*.

Shown below, even flat glass bends or 'lenses' light. Light from a point source is refracted when transmitted through a piece of flat glass. The slower speed of light in the glass causes the light waves to 'stack up' inside of glass. As with water, this bends or *lenses* the light. When light leaves the glass, it is bent again on the way out.

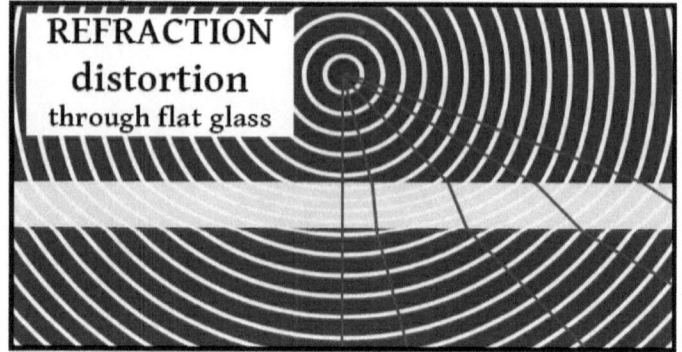

The change in the angle of the light wave may be seen by tracing the straight path of any one segment of a red line, and seeing where the light would *appear* to originate from as shown below. Also shown, these effects are easiest to detect when the source of the light is close to the surface of the glass.

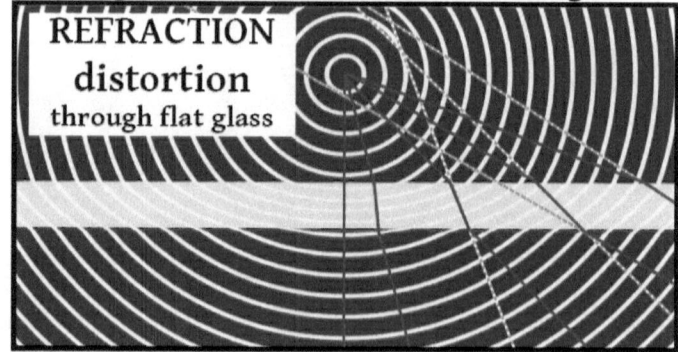

Irregular surfaces have unusual effects. A lens, shown below, may be used to refocus, concentrate, and/or dissipate light.

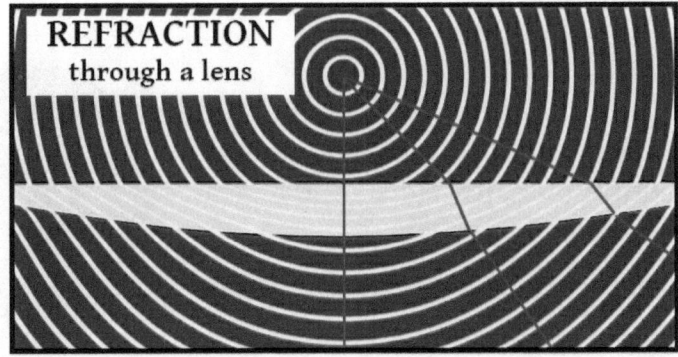

Prisms, shown below, spread light unevenly and that causes, among other things, waves of opposite peaks near the tip of the prism and a 'shadow'. This happens, in part, because opposite waves of light cancel each other out.

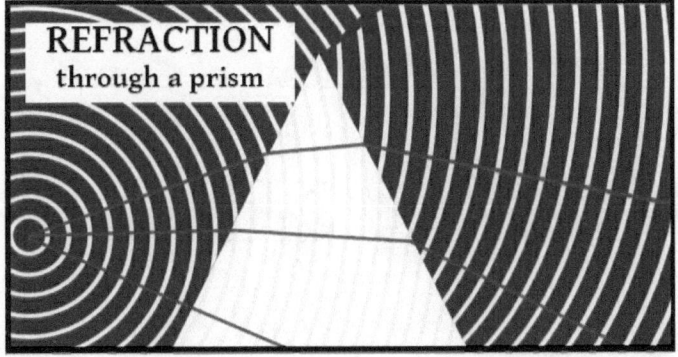

21

Galileo Galilei

Galileo Galilei (1564 – 1642) was an Italian mathematician, physicist, and engineer.

Galileo famously disproved that idea that heavy things fall faster.

Heavy things fall harder, but Galileo demonstrated that large metal balls fall at the <u>exact same speed</u> as small metal balls. This very important observation demonstrates that gravity accelerates all objects at the same rate. This may be thought of as the first law of gravity: Size does not matter.

First Law of Gravity

Size Does Not Matter

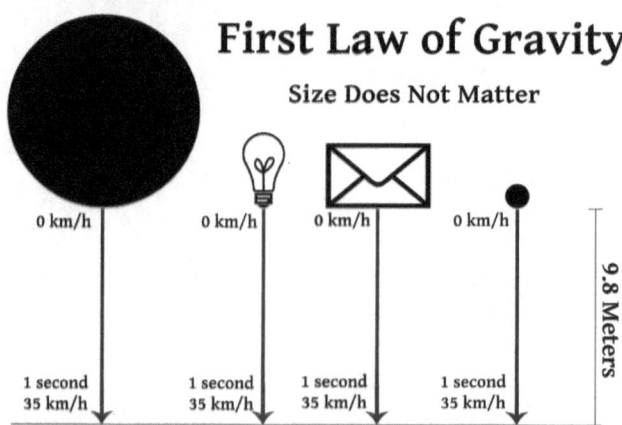

This principle was most famously demonstrated by Commander David Scott – an astronaut on the Apollo 15 – during a visit to the Moon on July 30, 1971.

Standing on the surface of the Moon, Commander Scott simultaneously dropped a hammer and a feather, and they hit the ground at the same time. Normally, meaning on Earth, Earth's atmosphere slows the feather down when falling, so hammers on Earth fall faster than feathers. On the Moon, the atmosphere is so thin that the hammer and the feather fell at about the same speed.

Galileo pioneered telescope technology, engineering and building many of his own devices. Employing these inventions as an astronomer, Galileo identified several of Jupiter's moons as orbital objects. From these and other observations, Galileo developed a revolutionary concept: Earth was not the center of the universe.

In Galileo's Italy, the overriding influence of the Catholic Church dictated both social and scientific philosophy. Accordingly, theories of creation were heavily influenced by theories of the heavens.

The Church's interpretation of ancient record was that God created the Heavens, the Earth, the Sun, the Moon, and the stars, and that he did so in that order.

With the location of Heaven unestablished, it stood to reason that Earth was the first material creation and was, therefore, the center of creation and the center of the universe.

Galileo lashed out at the Church and even wrote to the Pope directly, reprimanding him for the sermons about creation that he, Galileo – a devout Catholic – protested as false.

> *Freedom of Speech and Freedom of Religion are fundamentally important to the advancement of science.*

From Galileo's observations, the Earth was orbiting the Sun and not the other way around. The planets were also orbiting the Sun, and the stars were not orbiting anything.

For writing and teaching about these discoveries – or as revenge for reprimanding the Pope – Galileo was tried for heresy during the Roman Inquisition in 1615. The inquisition concluded that Galileo's heliocentric model was "foolish and absurd in philosophy, and formally heretical since it explicitly contradicts in many places the sense of Holy Scripture".

Galileo was convicted of violating "the sense of Holy Scripture".

For his politically incorrect opinions, Galileo was sentenced to death unless he publicly repented of his sinful teachings. Galileo withdrew his heliocentric model and apologized. He was then sentenced to life in prison, and after a few years in the dungeon, Galileo spent the rest of his life in house arrest.

Galileo spent most of his remaining life examining matters of physics which the Church would not kill him for, including studies of the structural integrity of various substances. Galileo's optical studies turned from the very far away to the very small, and he developed microscopes.

When Galileo grew old, he returned to writing and publishing concepts of astronomy. He did not, however, have much new information since the church did not allow him to own, possess, or use any telescope.

Robert Boyle (1627 – 1671) was an Anglo-Irish natural philosopher, physicist, inventor, and chemist. In 1641 Boyle was a 14-year-old boy who was in Italy with his tutor studying the "paradoxes of the great star-gazer," Galileo Galilei.

Twenty years later, Robert Boyle published *The Sceptical Chymist:* or *Chymico-Physical Doubts & Paradoxes (1661)*. Following up on Galileo's insights into the microscopic, Boyle studied how different chemical combinations have different attributes. His insights and theories ultimately established a new field of natural philosophy which we call *chemistry*.

Boyle is best known for *Boyle's Law* - the relationship of temperature and pressure as relating to gases. Boyle's law states that "for a fixed amount of an ideal gas kept at a fixed temperature, pressure and volume are inversely proportional."

Pressure * Volume = k (constant for the gas in question)

$$Pv = k$$

The use of the term *Boyle's Law* is contested, since Boyle was not necessarily the first person to figure this out. Boyle credited it to Richard Towneley (1629 – 1707) at Cambridge. Towneley had been assisted by Henry Power (1623 – 1668) at Cambridge, and some people attribute Boyle's Law to Henry Powers.

Christiaan Huygens

Christiaan Huygens (1629 – 1695), a Dutch physicist, mathematician, astronomer and inventor, was a major figure in the scientific revolution. Huygens is considered to be the first *theoretical physicist* and the key founder of *mathematical physics*.

In 1655 Huygens began grinding his own telescope lenses with the help of his brother. Back then, if you wanted better equipment, you had to build it.

In 1656 Huygens invented the *pendulum clock*. The pendulum clock was the most accurate non-solar method of time-keeping for the next three hundred years.

In 1657 Huygens published the first version of what would later become known as *Probability Theory* - a forerunner of the *Heisenberg Uncertainty Principle* - called *On Reasoning in Games of Chance.*

In 1659, when Isaac Newton was 18 years old, Huygens

derived the *Law of Centripetal Force* and published this discovery in *Di ve centrifuga.* Centripetal force, shown in the *illustration,* is the force which prevents object in orbit from falling out of orbit and going off in a straight line.

Centripetal force is used to calculate the force of gravity an orbital object experiences. Newton's *Principe* – with an advanced gravity equation – was published 28 years later in 1687.

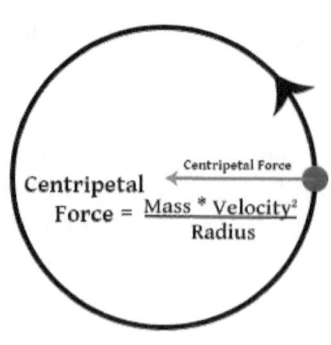

Centripetal Force

Centripetal Force = $\dfrac{Mass * Velocity^2}{Radius}$

In 1678 Huygens published *Traité de la Lumière* where Huygens showed how Snell's law of sines could be explained by, or derived from, the wave nature of light. This concept is now known as the *Huygens–Fresnel principle*.

Isaac Newton

Isaac Newton (1643-1727) was an English natural philosopher who defined the laws of motion. An avid astronomer, Newton precisely measured orbits and concluded that orbits were not deteriorating or slowing down.

From these observations, Newton concluded that objects in motion tend to stay in motion and that objects at rest tend to stay at rest. Isaac Newton compiled many of his findings in *Philosophiæ Naturalis Principia Mathematica* which was published in *1687*.

First law: In an inertial frame of reference, an object either remains at rest or continues to move at a constant velocity, unless acted upon by a force.

Force = mass

Second law: In an inertial frame of reference, the vector sum of the force(s) (F) on an object is equal to the mass (m) of that object multiplied by the acceleration (a) of the object.

Force = mass * acceleration

Third law: When one body exerts a force on a second body, the second body simultaneously exerts a force equal in magnitude and opposite in direction on the first body.

As shown in two graphics below, motion effectively moment-arily stops at the moment of contact. In that moment, the energy of motion has been transformed into magnetic potential. On an atomic level the positive (+) charges of atomic nuclei are repelling the incoming atomic nuclei (+). This creates equal and opposite force.

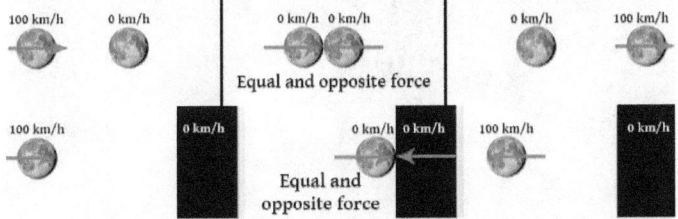

That magnetic repulsion then transforms into motion again, and energy is transferred or reversed.

The implied equation serves as the foundation of *thermodynamics* which was formed later.

$$\text{Force in} = \text{Force out}$$
$$(\text{input} = \text{output})$$

Of the three Laws of Motion, only the second one is typically represented with an equation.

Newton also established the Law of Gravity.

Law of Gravity:

$$\text{Force} = \text{gravity} * \text{mass} / \text{distance}^2$$

For two bodies

$$\text{Force} = \text{gravity} * (\text{mass}_1 * \text{mass}_2) / \text{distance}^2$$

Isaac Newton defined how to calculate gravity, but he never

found a method to determine the value of 'G' (gravity). Newton just proved that everything worked *proportionally.*

What Newton did <u>not</u> have was a mass value for the Earth, the Moon, or the Sun. Without one or more of those values, the value of G could only be guessed at.

Another significant problem is that gravity exerts *non-vertical* forces as illustrated in *Newton's Conundrum.* The values of non-vertical forces exerted by a body of *unknown consistency* (such as Earth) are very difficult to calculate, requiring highly specialized experiments.

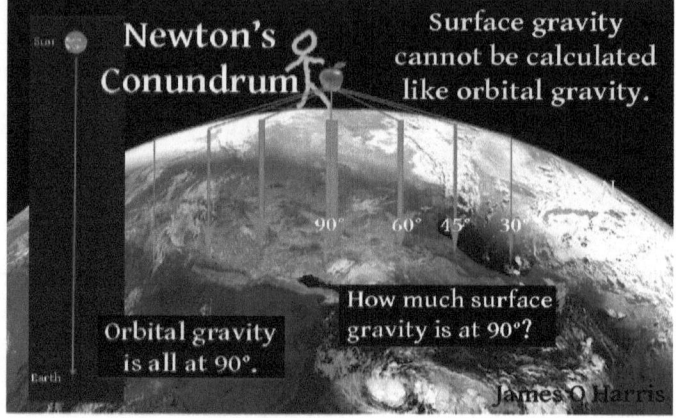

Isaac Newton was also disadvantaged by being stuck on Earth. He had very limited ability to experiment with gravity at differing altitudes. Newton wrote in *Principia* that he had formulated no theory concerning the origin or nature of gravity.

Newton's gravity, including gravity on Earth's surface, is still typically calculated as Copernicun Gravity.

Newton also heavily researched light and the nature of light. Newton met with Huygens during Huygens' visit to England in 1689. Huygen's concept of light differed from Newton's and Huygen's reasoning as to the wave-nature of light caused Isaac Newton significant when it came to publishing Newton's own theories.

In 1690, Huygens published *Treatise on Light* which is regarded as the first mathematical theory of light including the wave nature of light as defined in Snell's Law.

About that same time, Newton stated *Newton's Law of Cooling:* "The rate of heat loss of a body is directly proportional to the difference in the temperatures between the body and its surroundings." This would later be codified as *Fourier's Law.*

Opticks

In 1704 Newton published *Opticks: A Treatise of the Reflexions, Refractions, Inflexions and Colours of Light.*

Opticks introduced the color wheel and the concept of a prism as an object which divides light, shown in the illustration of a prism below.

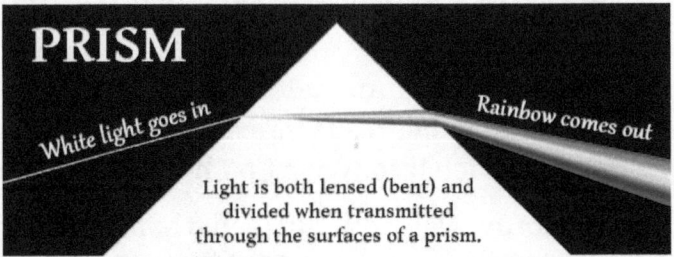

Light is both lensed (bent) and divided when transmitted through the surfaces of a prism.

Newton demonstrated that white light consists of many colors of light experienced at the same time. He suggested that prisms *divide* light into constituent, preexisting colors.

Newton demonstrated this to audiences using prisms but glass at that period in history was typically rather impure and imperfect. This caused some philosophers of the time to claim that the prism effect was false. Prisms were too 'unreliable' and some people said that Newton's results could not be replicated.

In the process of dividing and experimenting with light, Newton had developed the idea that light consisted of *corpuscles* – individual, independent particles of energetic nature or what we might now call *photons.* Newton reasoned that the various corpuscles traveled and that they interacted differently with matter, resulting in the prism effect.

The world of science now hinged on the assumptions of Copernicun gravity and that light consists of flying photons.

The model Newton espoused was that photons flew around got and got caught up in interactions with matter. Matter, he supposed, was like pudding. Photons flying through it would get caught the pudding from time to time, like plums, forming atomic plum pudding.

In this model, matter was no solid structure at all, but was more like a field. It was a field because gravity. If matter was a consolidated central unit, why would matter affect other matter as through gravity and magnetism? If matter was a centralized unit, why would it stick together. Why wouldn't it collide?

Newton's concept of the plum pudding atom justified the field attributes of matter and various interactions. The term 'plum pudding' emerged much later with J.J. Thomson.

Through astronomy Newton noticed that light seems to travel virtually infinite distances. For Newton's *'flying photons'* to complete such journeys, Newton reasoned that space must be composed of nothing.

Extending on the concept that space was composed of nothing, Newton observed that if space was nothing, then the laws of physics need not be cohesive from one place to another.

Being able to justify incongruent physics from one place to another, Newton originated a 'multi-verse' concept, where God might design things in any number of ways.

> *"And since Space is divisible in infinitum, and Matter is not necessarily in all places, it may be also allow'd that God is able to create Particles of Matter of several Sizes and Figures, and in several Proportions to Space, and*

perhaps of different Densities and Forces, and thereby to vary the Laws of Nature, and make Worlds of several sorts in several Parts of the Universe."

- Isaac Newton

The first theory of light as flying photons spawned the theory of empty space.

The theory of empty space allowed for objects divided by space to have no effect on each other.

That disconnectedness allowed Newton to conceive a multiverse where the laws of physics were not universal.

The world of science now hinged on the assumptions of Copernicun gravity, flying photons, and that space consists of nothing.

Modern observations of galaxies far, far away do not support Newton's theory of a multi-verse running 'alternative versions of physics' in other cosmic places.

That, then, retracing Newton's logic, indicates that space is not made of nothing. Space being not made of nothing, in turn, indicates that light in transmission does not consist of flying photons.

Cavendish Earth

Henry Cavendish (1731 – 1810) was born in Nice, which in now part of southern France, and lived most of his life in London, England where he made famous advancements in the fields of science, chemistry, and physics, especially concerning water, hydrogen, oxygen, and carbon dioxide.

Cavendish is best known for the Cavendish experiment. The image below, by Henry Cavendish, Cavendish depicted the operation of the device.

In 1797-1798, Cavendish used this device where two heavy metal balls (W) would demonstrate their gravitational attraction to two smaller metal balls (above B). Those forces could be measured and recorded through a telescope (far away so no air, ground, or gravity disturbance occurs) through the ports at (T).

Cavendish measured how much force (W) exerted on (B). He then calculated how much downward force Earth had on (W) and (B).

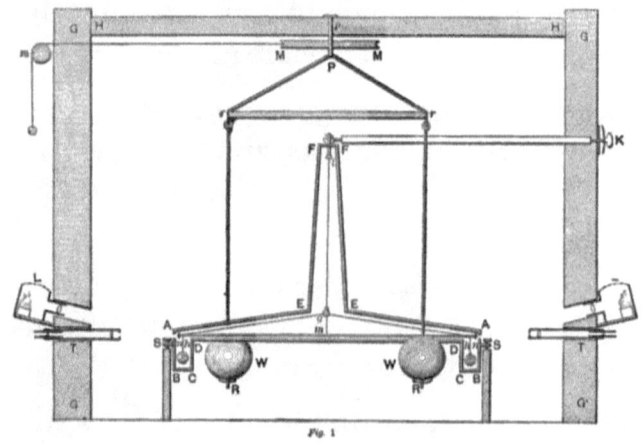

By Henry Cavendish - Cavendish,H.(1798), 'Experiments to determine the Density of the Earth' in McKenzie, A.S. ed. Scientific Memoirs Vol.9: The Laws of Gravitation, American Book Co. 1900, p.62

If (W) mass has G effect on (B) mass at distance r, and Earth mass has G effect on (W) mass at distance r^2, what is the value of G?
Seems easy enough to solve if all interactions are vertical.

And Cavendish solved this as indicating that the average density of Earth was 5.4 grams per centimeter^3, or, as he termed it, 5.4 times heavier than water.

That density is nearly twice the average density of Earth's surface material, which is about 2.79 g/cm^3. Cavendish therefore proposed that Earth's core was made of iron.

Iron weighs 7.8 g/cm³ solid and 6.98 g/cm³ as a liquid, so a big iron core could make Earth 5.4 g/cm³ on average. The concept that Earth is very heavy in the middle is the Cavendish Earth model.

The world of science now hinged on the assumptions of Copernicun gravity, flying photons, and a Cavendish Earth with a big, heavy, iron core.

Earth is presently valued as weighing 5.51g/cm³, making (theoretically) the densest large object in the Solar System.

Cavendish stressed that his experiment did not establish the value of G. He was right in doing so. Weight and mass are not the same thing. As seen in *Underestimating Gravity* below, Cavendish was not working with the full force of gravity in his observation.

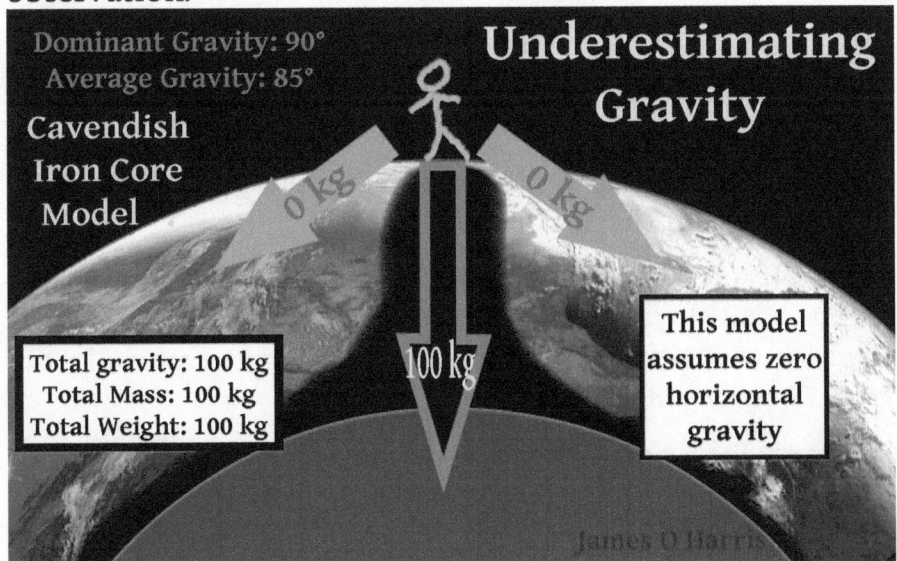

Cavendish may have been aware that. He understood some of the fundamental difficulties Newton faced. But Cavendish did not see far enough into this problem to realize that it falsifies his results as well. If less than 100% of surface gravity is vertical, then the weight of an object must be less than the gravity exerted on that object.

How much less?

The key to that puzzle is the angle of gravity – a phenomenon not recognized until two hundred years later.
Lacking that, the scientific community agreed that 5.4 g/cm^3 the Cavendish Earth model was accurate.

The world of science now hinged on the assumptions of Copernicun gravity, flying photons, Cavendish Earth, and more Copernicun gravity.

With an average value of Earth's density, a value of Earth's total mass was then easily calculated. From that, the value of G was clearly inferred as 6.67E-11. The value of 'G' is also known as the *Newtonian constant*.

The world of science now hinged on the assumptions of Copernicun gravity, flying photons, Cavendish Earth, gravity equals mass, and a Copernicun G value of 6.67E-11.

Cavendish found that *if* the Earth's angle of gravity was 90º, then the mass of Earth could be deduced.

In 1785 the Dutch physicist and chemist Martinus van Marum (1750 – 1837) was famous for his public demonstrations of chemical reactions and electrical discharges, especially his *Large Electricity Machine.*

The Large Electricity Machine was an electrical generator – in this case, an *electrostatic generator* or simply a big static electricity generator.

Marum's investigations into electricity and chemistry included early electrolysis of water and the detection of 'the smell' of high voltage – 'the smell' of ozone.

As with all inventions, discoveries, and ideas, there existed other people elsewhere doing things the world has never heard about. Whenever I say someone was 'first', I am not pretending that it never could have happened before.

1800 - 1850

Waves of Light

In 1801, Thomas Young (1773 – 1829) published *On the Theory of Light and Colours* which was backed up with a *single-slit* experiment which was later doubled as the double-slit experiment. Concerning this work, Thomsas said,

> "A very extensive class of phenomena leads us still more directly to the same conclusion; they consist chiefly of the production of colours by means of transparent plates, and by diffraction or inflection, none of which have been explained upon the supposition of emanation, in a manner sufficiently minute or comprehensive to satisfy the most candid even of the advocates for the projectile system; while on the other hand all of them may be at once understood, from the effect of the interference of double lights, in a manner nearly similar to that which constitutes in sound the sensation of a beat, when two strings forming an imperfect unison, are heard to vibrate together."

Simply put, Thomas said 'the particle theory of light transmission does not explain the many wave-like attributes of light transmission,' and, 'many of the wave properties of light correlate to wave properties of sound'

Single-Slit Experiment

In the illustration below, one high-energy photon is released from matter as a gamma-ray into a slit in a dark box. That energy spreads as waves and many photons occur on the surface inside of the box.

Photons exist only at the sides of the box, where light-waves interact with matter. Since these waves are constantly traveling, a wide area inside of the box is dimly lit. If time were stopped, the light would appear like an interference pattern as shown below.

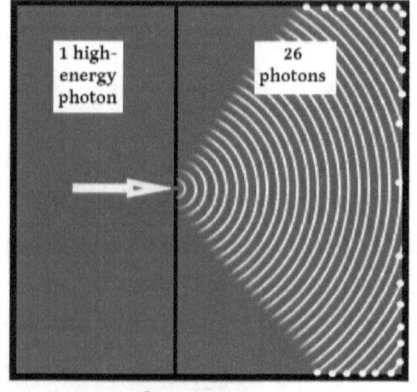

Single-Slit Experiment
Single-slit wall

1 high-energy photon

26 photons

Key Implications
· Light travels as waves
· The path of light is not straight
· Photons are divisible units
· Photons are not light

The double-slit experiment takes the single-slit experiment a step further.

Double-Slit Experiment

In the illustrated 'Wave Pattern' box the gamma ray spreads as waves. Each light wave represents one gamma-ray photon charge. Since the waves are spreading the energy of the gamma-ray is distributed.

When those rays or waves affect matter, photons are induced. Photons are induced wherever a sufficiently powerful light wave encounters matter.

The sides of the box are composed of matter, so that is where photons occur.

Double-Slit Experiment

Single-slit wall

Double-slit wall

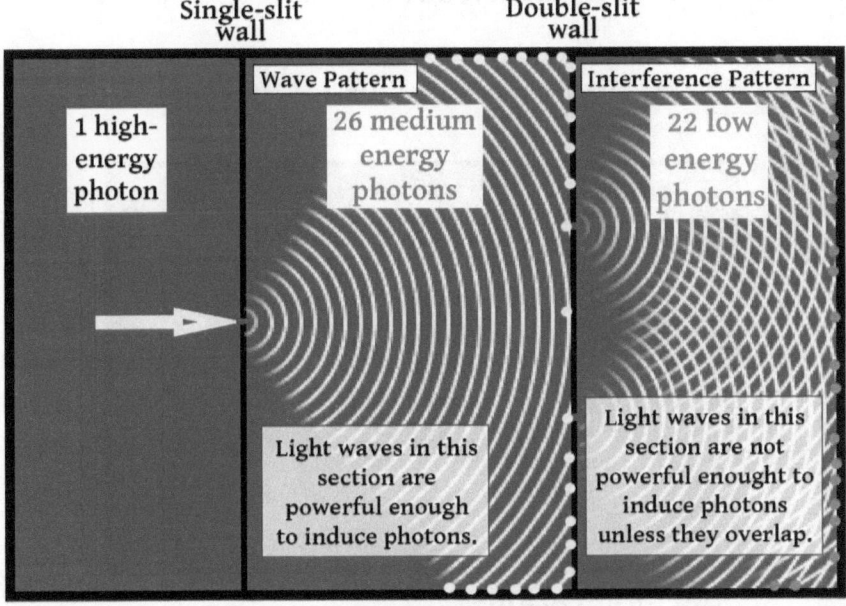

Wave Pattern

1 high-energy photon

26 medium energy photons

Interference Pattern

22 low energy photons

Light waves in this section are powerful enough to induce photons.

Light waves in this section are not powerful enought to induce photons unless they overlap.

Key Implications

- Light travels as waves
- The path of light is not straight
- Photons are divisible units
- Photons do not exist in transmission

In the 'Interference Pattern' box the light waves are much weaker, so a photon will only occur wherever the light waves cross and, at the same time, encounter matter. Light waves there are too weak to induce photons unless the light waves are combined or intersecting.

To see how the double-slit experiment conflicts with flying photon theory, consider what would happen if light *in transmission* consisted of photons. As seen below, if light in transmission consisted of photons, there would not be any wave pattern and there would not be any interference pattern.

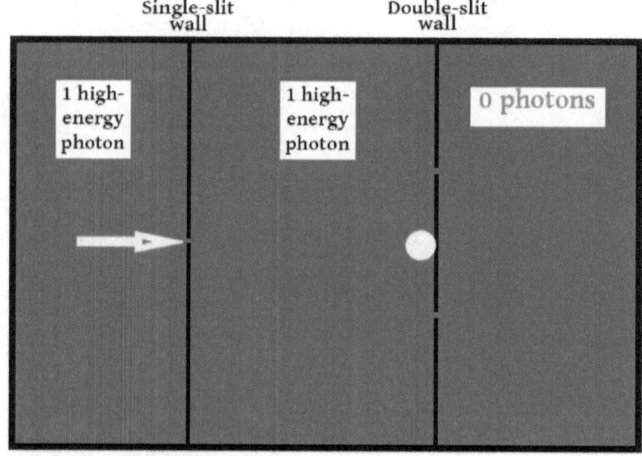

Flying Photon Model and Expectations

Single-slit wall   Double-slit wall

| 1 high-energy photon | 1 high-energy photon | 0 photons |

Key Implications

- Light travels as photons
- The path of light is straight
- Photons are indivisible units
- Photons transmit light

In 1814 Josef von Fraunhofer (1787 – 1826), a Bavarian physicist and optical lens manufacturer, built the first spectroscope and used it to observe the Sun. He discovered that sunlight was not white, even light, but that sunlight, when divided into its component spectra. It had dark regions. Certain wavelengths of light were missing.

Arago Spot

In 1818 the French Academy of Sciences (1666 -), now part of the Institut de France (1785 -), held a competition to explain the properties of light. Augustin-Jean Fresnel (French 1788 – 1872) proposed that light travels as waves. Dominique François Jean Arago (French 1786 – 1853) proposed that light travels as particles.

Arago proposed an experiment to prove whether light was a wave. A point source of light would be shined on disc two millimeters in diameter. If any light showed up behind the disk, light was a wave. If no light showed up behind the disk, light was a particle.

The experiment took place during the competition. A spot in

the middle of the shadow of the disk showed up and was called the Arago spot, as shown. Aragos discovery of the Arago spot was taken to prove that light travels as waves, allowing Fresnel to win the competition.

In 1833, William Whewell (1794 – 1866) coined the term *scientist.* Whewell was an English polymath, scientist, Anglican priest, philosopher, theologian, and historian of science. Whewell was also a Master at Trinity College at Cambridge University.

Whewell died at Cambridge when he fell from his horse.

In 1833 Robert Wilhelm Bunsen (1811 – 1899) – the patron saint of pyromaniacs – demonstrated the use of iron oxide hydrate as the most effective antidote against arsenic poisoning.

Like many other scientists, Robert Bunsen, is not best known for his best work. Bunsen probably dedicated less than three days of his time to the creation of the Bunsen Burner. William Whewell is most famous for inventing the word 'scientist'. His second most famous achievement was falling from his horse.

Thermodynamics

In 1834 Benoît Paul Émile Clapeyron (1799 – 1864), a French a engineer, physicist, and founder of *thermodynamics*, published the *Ideal Gas Law.*

Pressure * Volume = n (moles of gas) * R (*ideal gas constant*) * Temperature

$$PV = nRT$$

The problem with the ideal gas law is that it changes with altitude. The ideal gas constant also differs for every known planet, moon, and star. The root of this problem is that <u>pressure is proportional to gravity</u>.

That means that to get the ideal gas constant to work right, a proper value of gravity and a proper understanding of planetary structure must exist.

As demonstrated (formerly) in *Underestimating Gravity*, misunderstandings concerning mass/weight proportionality and planetary structure have prevented the vale of G from being established effectively. Since the force of gravity was not effectively established, the ideal gas law has always seemed flawed or conditional. A proper valuation of G should resolve these problems.

The world of science now hinged on the assumptions of Copernicun gravity, flying photons, Cavendish Earth, gravity equality, 6.67, and that the ideal gas law does not actually have real constant.

In 1834, Clapeyron published *Memoir on the Motive Power of Heat.* This paper predicted that heat could be used to power an engine – the way that modern internal combustion engines work.

'The Smell' of Electrolysis

In 1839 Christian Friedrich Schönbein (1799 – 1868) identified 'the smell'. The 'smell' occurred in the vicinity of electrical discharges and lightening strikes and had been scientifically pondered since before 1785.

While working with electrolysis of water, Schönbein became aware of 'the smell' and coined the Greek-language term of 'the smell' to identify it. The Greek word for 'the smell' is 'ozein' and the gas is now commonly called 'ozone'.

Ozone consists of three oxygen atoms existing as one molecule or *allotrope (molecule consisting of one element)*. Atmospheric oxygen is also an oxygen allotrope and consists of two oxygen atoms. Ozone is a highly unstable, extremely reactive, and highly acidic. Ozone is one of the most powerful oxidizers known.

Ozone has a very short half life in caustic environments, in positively-biased environments, and in warm environments. Ozone is negatively charged.

Schönbein encountered ozone when working with water. Water goes through electrolysis when electrical voltage is applied. In the illustration below, oxygen atoms are attracted to the positive (+) electrical post or *anode*, while hydrogen atoms are attracted to the negative (-) electrical post or *cathode*.

Ozone forms near the cathode (-) but does not persist in positively-charged environments such as among the oxygen gathered by the anode (+) during electrolysis.

Electrolysis of Water

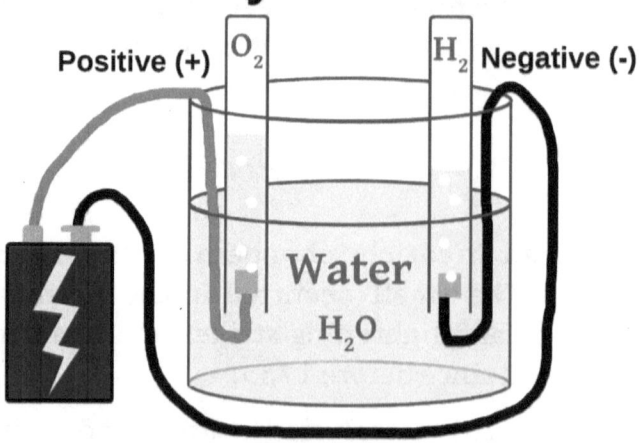

Electrolysis chemically separates
water into oxygen and hydrogen.

Ozone occurs where an abundance of negative (-) electrical energy exists.

- In the upper atmosphere, where ultraviolet radiation is intense,
- Near lightning strikes,
- In the vicinity of high voltage electrical discharges,
- When rain falls, the motion of the falling water causes

electrical inductions within the atmosphere – static electricity – and that also generates ozone.

In the electrolysis experiment, ozone will form dissolved in the water near the cathode (-) and will then attract to the anode (+). Near the anode, the positive (+) electrical field re-forms ozone into atmospheric oxygen.

Faraday, Induction, and Electrical Motors

In 1839 Michael Faraday (1791 – 1867) laid the foundations for the modern sciences of electrodynamics, electromagnet-ism, and induction. The law of induction became known as Faraday's Law.
Induction occurs when one object moves past another object. When that happens, the electromagnetic fields interact. If that interaction is strong, as with magnets moving past each other, then electricity is produced.

How fortunate for civilization, that Beethoven, Mich-elangelo, Galileo and Faraday were not required by law to attend schools where their total personalities would have been operated upon to make them learn acceptable ways of participating as members of "the group".

- Joel H. Hildebrand's Education for Creativity in the Sci-ences speech at New York University, 1963

Faraday built the first known electromagnetic motors – a homo-polar generator and a dynamo – devices presently known as 'generators' and 'DC motors'. The core principles of all motors, generators, and alternators relate to rotational induction.

In the illustration below, the copper electromagnets of the motor attract to the stationary magnets mounted to the stator. The stator is the body and permanent magnets which remain stationary. The attraction causes the rotor to rotate. As the rotor rotates, the electromagnets disconnect from the power source at the commutator. The electromagnet goes neutral and the rotor continues to spin. The spinning commutator then reconnects to the power source and charges the electromagnets again, causing the motor to continue to spin. In the example given, the polarity of the electromagnet reverses once per rotation, so the negatively charged pole of the electromagnet is always causing force in the same direction – clockwise.

Rotational Induction

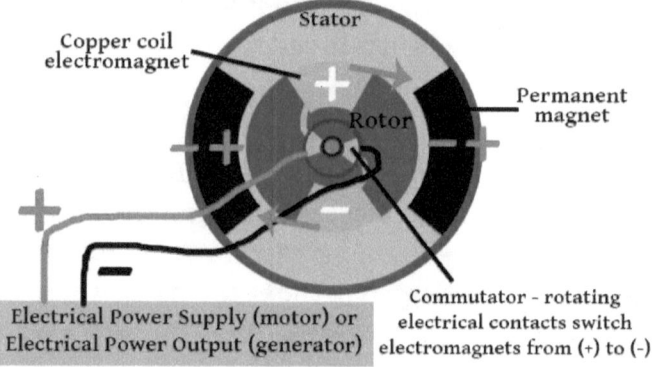

Copper coil electromagnet
Stator
Rotor
Permanent magnet

Electrical Power Supply (motor) or
Electrical Power Output (generator)

Commutator - rotating electrical contacts switch electromagnets from (+) to (-)

Rotational induction can also be reversed, applying motion to spin the rotor and collecting electricity from the motor, as a generator instead.

Faraday also demonstrated that electromagnetic induction could be interrupted by enclosing an object within a metal pail or screen, an arrangement now known as a *Faraday cage*.

For a Faraday cage to be effective, the openings (if it is a screen like a microwave oven door) must be less than one half of the length of the wavelength to be blocked.

If a cellular phone is wrapped in aluminum foil (or any metal), it will not receive phone calls or connect to any wireless network. The network will not see the phone, either. The aluminum foil forms a Faraday cage which blocks the radio waves.

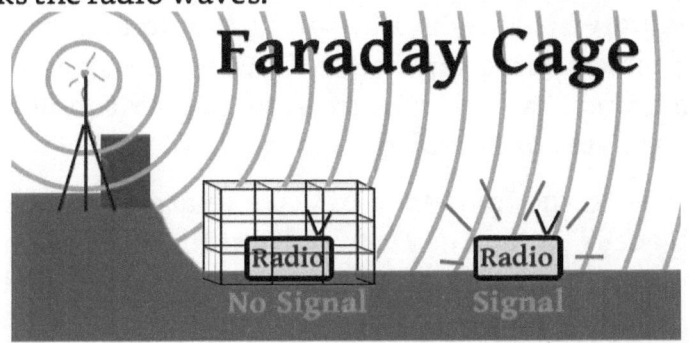

Faraday cages evidence that light is transmitted as waves. In the illustration above, for the Faraday cage to block radio-waves of 100 MHz, the cage may have gaps as large as 16.7 centimeters.

Electrons have a radius of about 2.82E-13 centimeters. About 7.21E+13 electrons can fit through the identified Faraday cage side by side through every opening. But no signal gets through. No electrons from the radio tower are detected inside the Faraday cage. This evidences that light and radio travel as waves – that photons and electrons do not exist in transmission. If they did, electromagnetic radiation would not be stopped by a Faraday cage.

This is a field reaction.

Near the end of 1839 Faraday had a total emotional break-down. Following eight years of intense experimental investigation, Faraday publicized his concepts and was then barely seen again for six years. Faraday's later year were spent in search of a unified theory – to define gravity as an electromagnetic force

The Doppler Effect

In 1842, Christian Doppler (1803-1853), an Austrian mathematician, physicist, and astronomer, published *On the coloured light of the binary stars and some other stars of the heavens.* Doppler postulated that light, like sound, would become dilated or compressed based on the motion of the object emitting that light, and that this change in frequency can be used to measure the speeds of orbits.

The change to the frequency of wave relative to the motion of an object is called the *Doppler Effect* for local phenomena or *Dopplerian Red-shift* when referring to astrological observations.

The Doppler Effect can be observed with sound waves. If an object is making a constant sound, and if the air conducting that sound is not moving, an object will sound 'normal' as shown.

If that same object is then thrown away from the observer, and the air is not moving, that object's sound becomes dilated and arrives at the observer with longer intervals between the waves. The sound becomes shifted to a lower frequency. It becomes red-shifted as diagrammed below.

When an object is making a sound and moving away from you, the sound you hear will be a lower pitch than the sound the object is actually making.

The opposite is also true. If the object is thrown *at* the observer, and the air is not moving, then the timing of the sound waves is compressed. The sound will arrive with shorter intervals between the waves. The sound becomes shifted to a higher frequency. It becomes blue-shifted.

When an object is making a sound and moving towards you, the sound you hear will be a higher pitch than the sound the object is actually making as diagrammed.

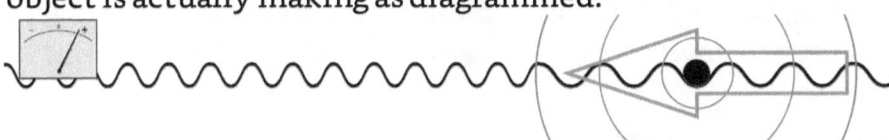

And if something like Chuck Yeager is coming towards you faster than the speed of sound, you will not hear him coming at all. He will hit you directly in the face before the sound arrives. Sound does not travel faster than sound, as represented below.

Sound and radio wave speeds are also affected by changes in the density of atmosphere. Changes in the speed of waves due to atmospheric density is also called the *Doppler Effect.* The heavier the atmosphere is, the slower the radio waves broadcast through it.

The Doppler Effect is popularly used in weather detection technology and movie theater sound technologies.

In 1845 Gustav Robert Kirchhoff (1824 – 1887) formulated the *circuit laws* still used today in electrical engineering. Kirchhoff was a German physicist and a pioneer in the fields of electricity, spectroscopy, and black body radiation

In 1848, the *February Revolution* occurred in France. The February Revolution established the principle of *right to work* in France – a concept not allowed under monarchies or fascist governments. *Right to work* is fundamentally important to freedom of expression, freedom of speech, and economic freedom.

Origin of Species

In 1849, Charles Robert Darwin (1809 – 1882) published a book that rocked the scientific world – *The Origin of Species.* Darwin posited that the differentiation of creatures on Earth was due to environmental pressures and subtle genetic mutations.

As a geologist, Darwin pushed creation theory back 300 million years. Observing the southern region of England called the *Wealds and Downs*, Darwin posited that the period of time it would take for erosion to carve out that region at the present rate would be 300 million years. Since Earth had to form before that began, the Earth's age was speculated back to 400 million years ago.

That greatly conflicted with contemporary views ranging from 6,000 years to 20 million years.

The concept of evolution entered the scientific mind and re-arranged every concept of Earth and its creation.

Instead of God breathing life into some dirt, lightning struck some mud.

This dramatic shift in thinking ultimately evicted the Bible from the science classroom.

From plate tectonics to big bangs to radioactive substances, Darwin's timeline became the new gold standard for measuring the age of the Earth and everything in it. Subsequent fossil finds would push the estimate of Earth's age back much further.

The world of science now hinged on the assumptions of Copernicun gravity, flying photons, Cavendish Earth, grav-

ity equality, 6.67, strange gases, that life-forms are random chemical accidents, and that evolutionary fossils are the key to determining the age of the Earth.

While Darwin was digging up his family tree, Foucault proved that the Earth spins.

In 1849 Léon Foucault (1819-1868) experimentally demonstrated that absorption and emission lines appearing at the same wavelength are both due to the same material. This led Foucault to develop the science of spectroscopy. Foucault also recognized that temperature affected where those emission lines appeared on the spectrum.

1850 - 1900

In 1851 Foucault provided the first direct (non-astronomical) evidence of Earth's rotation. Essentially, he just set up a really big pendulum in the Panthéon in Paris, France.

Due to the force of Earth's spin and gravitational forces, the pendulum did not point straight. That is not immediately obvious. The pendulum works the same way common bubble-level works. Gravitationally, the pendulum is always vertical.

However, if you watch it for 24 hours, the pendulum will be observed making precessional circles relative to Earth's rotation. The effect is greatly exaggerated if the pendulum has been set in motion and is swinging.

When used to measure these effects, the device is called *Foucault's Pendulum.*

The farther that Foucault's pendulum is from the equator,

the more exaggerated the pendulum's angle will be and larger the circles. On the the equator, Foucault's pendulum does not move or change.

This effect becomes very important when engineering very tall buildings. These shifting forces can destroy a poorly engineered building, bridge, or damn.

Foucault's talent for physics extended beyond giant pendulums. In the field of electricity, Foucault discovered *eddy currents*. In the field of physics, Foucault also coined the term for the 'gyroscope'.

In 1851, Foucault "drove the last nail into the coffin" of Newton's *corpuscular theory of light* with the demonstration of the *Fizeau–Foucault apparatus*.

The device measured the speed of light with a mirror several kilometers away, directing light at it through a spinning gear. That light would then travel to the distant mirror and be reflected back through the gear. When the gear was sitting still, a certain amount of light would make it back through. But when the gear was set in motion, less light made it through. That meant that light had a limited speed.

Newton's light had unlimited speed.

But if the speed of light were infinite, the same amount of light would make it back through the gear no matter how fast the gear was spinning.

In 1851, Armand Hippolyte Louis Fizeau (1819 – 1896), a French physicist and associate of Foucault, measured the speed of light as being 313,300 km/s using the Fizeau–Foucault apparatus.

Fizeau then calculated how fast the speed of light must be based on how much shadow he got. Fizeau calculated the speed of light to within 5%.

Fizeau's accomplishments also included the invention of the

electrical *capacitor* – a device capable of holding and storing electrical charges. Unlike a battery, a capacitor does not use chemical reactions to hold charges.

Faraday's Folly

In 1855, Faraday was going after gravity. Gravity, Faraday insisted, was a magnetic force and he went after it with experimental investigation.

Gravity is an active and observable force, a force of limited potential, a force which may be converted into electricity (as through the use of a hydro-electric dam), and a force which is affected by and/or overpowered by magnetism.

Is gravity an electromagnetic force?

Unlike other forces, gravity and magnetism do not diminish, spread, reduce, or entropy. The Earth has orbited the Sun for billions of years and the force of gravity has not diminished at all.

Light waves are energetic disturbances which are broadcast outwards at the speed of light. Those disturbances continue outwards at the speed of light until disturbed by other material interaction. The forces of these interactions are counted in *photon-volts* or *electron-volts*.

Photon: A particle having either a positive or negative charge; quanta
Photon-volt: The smallest discernible unit of light energy; quanta.
Quanta: The smallest discernible value.
Electron: A cloud of about 2,315 unique negatively-charged particles/photons.
Electron-volt: The difference of potential of one electron; about 1/2,230th of one volt.
Difference of Potential: Energy which would be expressed if

released or grounded.
Volt: The difference of potential of one coulomb; about 2,230 electron volts; about 5.2 million photon volts.
Watt: One coulomb per second.

Wattage can be extended to include *horse-power* or raw physical force, generally. Thus, all forces may be defined in electrodynamic terminology, and the force of gravity may also be expressed in watts.

Light spreads, sound-waves spread, gravity waves spread, but gravity, mass, and magnetism stay right where they are.

How can all of these phenomena express wattage, but only half of them require any input? What is the input that powers gravity? Where is the input wattage?

Einstein, later, ran into this problem as well. Einstein's solution was to consider gravity as *not a force.* That way, no input is needed. Of course, that completely fails to address the obvious output that exists, as through hydro-electric dams and the Earth not falling out of orbit.

Hydro-electric dams are powered by gravity. Some people mistakenly believe that they are powered by the Sun via evaporation cycles. If there were no gravity, all evaporation would be just that: evaporation. Without gravity there would not be condensation, rainfall, or hydro-electric damns.

So what's powering all of that?

I submit that any theory which cannot account for this source can never succeed as a universal or unified theory. Any theory which does not account for this does not address the most fundamental of physical laws: Input equals output. Hydro-electric dams generate output. What is the input?

The answer to the riddle that drove Faraday mad is *Atomic Energy Transformation (AET).* AET is detailed in Part II: Paradox Null.

So what happens if I try to convert mass to energy?

Modern high-energy sciences have answered that question.

Every light element is endothermic when fissioned, and (theoretically) exothermic when fused.

Every heavy element is exothermic when fissioned, and (theoretically) endothermic when fused.

If you take equally weighted units of every element on the periodic chart and fuse them all together, the total energetic output will be zero.

If you take equally weighted units of every element on the periodic chart and fission them all apart, the total energetic output would be zero.

If you fissioned or fused all of the atomic material of planet Earth, the net output would be zero.

The net average mass/energy conversion ratio is 0/0.

What did Faraday get when he tried? He got nothing. <u>Faraday's null result of converting mass to energy was accurate.</u>

Faraday submitted his results to the Royal Society, but the Royal Society refused to publish his null result.

Faraday was challenging one of the most fundamental forces of nature. When Faraday suggested that gravity was not an intrinsic force, but one that could be manipulated, he was seen as contradicting the great Isaac Newton himself. Faraday was then shunned by the scientific community.

Unable to discount Faraday's many brilliant findings and assertions, the scientific community retained its acceptance of Faraday's early findings and condemned his later assertions as lunacy. This was justified by claiming that Faraday had become senile.

"To achieve anything really worthwhile in research it is

necessary to go against the opinions of one's fellows. To do so successfully, not merely becoming a crackpot, requires fine judgment, especially on long-term issues that cannot be settled quickly. ...To hold popular opinion is cheap, costing nothing in reputation, whereas to accept that there is evidence pointing oppositely... is to risk scientific tar and feathers. Yet not to take the risk is to make certain that, if something new is really there, you won't be the one to find it."

- Fred Hoyle, Home Is Where the Wind Blows:

Chapters from a Cosmologist's Life (1994) p. 235

Faraday passed away without completing his quest for a universal theory.

The scientifically toxic assumption which was attached to Faraday's folly was that gravity was a force independent of magnetism, and not an electromagnetic or proto-electromagnetic effect.

The world of science now hinged on the assumptions of Copernicun gravity, flying photons, Cavendish Earth, gravity equality, 6.67, strange gases, random evolution, Darwin's timeline, and the notion that gravity has nothing to do with any other natural force.

Spectroscopy and Playing with Fire

In 1854 the University of Heidelberg in Germany was building Robert Bunsen a laboratory and Bunsen called Gustav Kirchhoff to work with him there. Kirchhoff accepted.

Interested in keeping his laboratory clean, Bunsen suggested

to Peter Desaga – the university's mech-
anical engineer – a design for a valve-con-
trolled gas burner which mixed gas with
air before allowing it to burn. Mixing the
gas with air before letting it burn created
a clean, sootless flame. Laboratories of
that time period were notoriously dirty –
being blackened with soot generated by
burning coal gas or other fuels with
standard burners.

Premixing the fuel and air significantly reduced the soot generated and the valve allowed for control over the size of the
flame.

The Bunsen burner was not a
patent-new idea and Bunsen
didn't patent it. Such a design
had been used formerly by
Michael Faraday. Bunsen burners resembled a design patented in 1856 by German gas
engineer R. W. Elsner. The Bun-

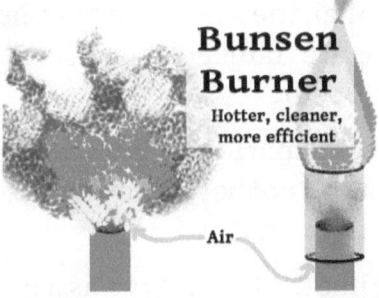

sen-model Bunsen burner was not patented and the design
spread like wildfire throughout academia.

In 1857, Kirchhoff calculated that the speed of electricity in a
zero-resistance wire would be equal to the speed of light. This
evidenced the connectedness of light and electromagnetic
phenomena.

In 1859, Kirchhoff and Bunsen invented and constructed
spectroscopes. Using Bunsen's sootless burners, Kirchhoff and
Bunsen experimented with all manner of substances, burning
them with the Bunsen burner and observing their spectroscopic output. All in the name of science.

Kirchhoff's Laws of Spectroscopy

From Kirchhoff's and Bunsen's observations Kirchhoff created the *laws of spectroscopy.*

Spectroscopy is the study of how different materials – particularly gases – react and respond to light based on the spectra emitted and/or absorbed. Kirchhoff's laws demonstrate that electromagnetic phenomena affect all manner of atomic structures.

This concept is foundational to *Atomic Energy Transformation (AET)*, which is covered later in this book. When an atomic structure is *affected* by any electromagnetic, magnetic, gravitational, inertial, or other phenomenon, photons and/or electrons are induced and destroyed and/or exchanges of momentum occur.

Transformation occurs whenever energy changes from one form to another or momentum has been transferred from one object to another.

Kirchhoff's Laws serve as a guide for some of the laws governing this transformation.

Kirchhoff's First Law – Continuous Emission Spectra

This law demonstrates that when a dense atomic substance is highly energetic and very hot, that that substance will emit a full range of light – white light.

It is interesting that incoming frequency is not particularly important. No matter what frequency matter is over-charged with, it will redistribute that frequency across a whole range of frequencies when highly excited.

Kirchhoff's First Law of Spectroscopy

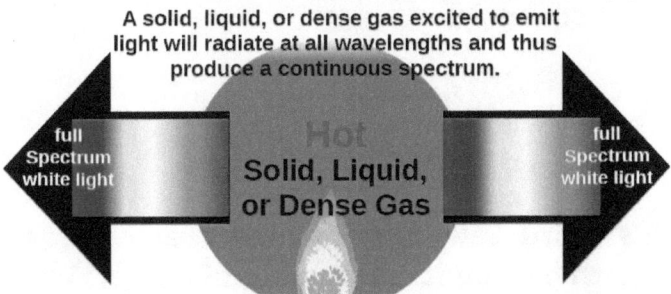

A solid, liquid, or dense gas excited to emit light will radiate at all wavelengths and thus produce a continuous spectrum.

When the photon population gets too high, atoms release those energies at any available frequency, resulting a full spectrum or *white light*.

Kirchhoff's Second Law – Emission Spectra

Kirchhoff's Second Law demonstrates that different atomic structures have preferential discharge frequencies based on unique atomic structures. When not crowded together, these highly charged atoms absorb the vast majority of frequencies, and re-emit only their own signature frequencies.

This is the *emission spectrum*. This is the music of the

Kirchhoff's Second Law of Spectroscopy

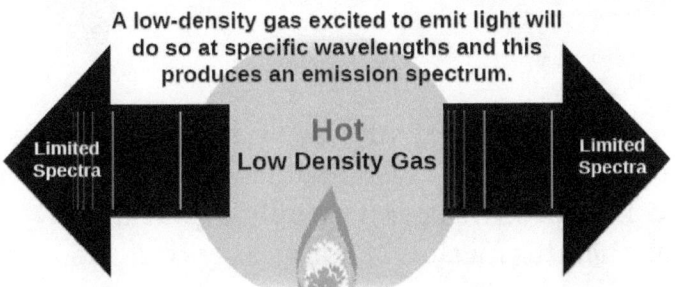

A low-density gas excited to emit light will do so at specific wavelengths and this produces an emission spectrum.

atom. Harmonic frequencies are bright. Resonant frequencies are absorbed. Most frequencies are resonant, so most of the spectrum is black. Resonant-frequency energy has been absorbed and transformed into harmonic frequencies. Most of the energy, however, is transformed down to infrared or microwave frequencies.

Kirchhoff's Third Law – Absorption Spectra

Kirchhoff's Third Law demonstrates matter most readily interacts with harmonic frequencies of energy. When not in a highly excited state – and when photon shells of atoms are much smaller – resonant frequencies do not interact with matter.

In cool, low density situations, harmonic frequencies are absorbed and those are transformed into lower octaves of frequency – to energies including infrared and microwave frequencies. This *removes* the harmonic frequencies of light from the visible spectrum resulting in an *absorption spectrum* which can be used to identify substances.

Kirchhoff's Third Law of Spectroscopy

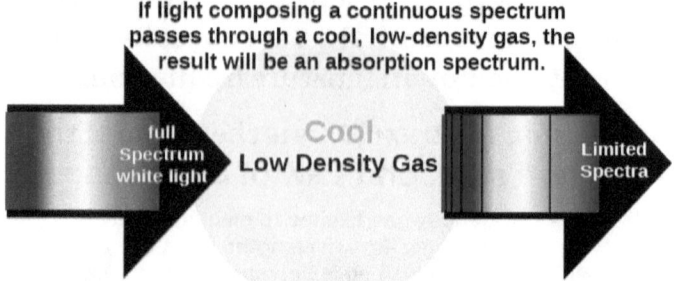

If light composing a continuous spectrum passes through a cool, low-density gas, the result will be an absorption spectrum.

full Spectrum white light

Cool Low Density Gas

Limited Spectra

Since the emission spectrum is the same as the absorption spectrum, astronomers must decide whether an object is cold or hot to interpret its spectroscopic signatures. We will see later that that distinction is not as easy to make as formerly thought.

In 1859, Kirchhoff used spectroscopic data to prove the presence of sodium on the Sun. Sodium is about three hundred times heavier than hydrogen.

Also in 1859, Kirchhoff proposed the *law of thermal radiation*. He provided proofs for the *law of thermal radiation* two years later in 1861. The law of thermal radiation observes that heat is not directionally preferential. It just 'flows' from hot to cold. This is the concept underlying black body radiation which was defined later.

Solar Disposition

1n 1859, Richard Christopher Carrington (1826 – 1875) – an amateur English astronomer reported the discovery of solar flares and along with that, the discovery of the solar wind – a stream of gases flowing out and away from the Sun.

For the Sun to throw off such large amounts of mass – solar flares and *coronal mass ejections* – contradicted conventional wisdom. For such ejections to occur, the pressure on the surface of the Sun had to be exponentially lower than formerly assumed or matter could not leave the Sun and no solar wind could exist.

Carrington used sun spots to gauge the rotation of the Sun, which he found to be 25.38 days within 45° of the equator. But at higher latitudes, Carrington could not find anything to measure rotation by.

Continuing the burner experiments at Heidelberg University, Kirchhoff and Bunsen discovered cesium in 1860 and discovered rubidium in 1861.

The difference in the angle of light between the first and last leg of travel across this arrangement allows the speed of light to be observed.

In 1861 William Crookes (1832 - 1919), using the spectroscopy techniques of Kirchhoff and Bunsen, identified a new element he called *thallium*. Crookes most famously invented the *Crookes Tube* which demonstrated that the path of light could be manipulated using magnetic or electromagnetic impulses.

A forerunner to the Cathode Tube, these concepts will be covered with Cathode tubes later.

In 1862 Foucault gave his final word as to the speed of light: 299,796 kilometers per second. He was off by merely 0.00118%. Foucault's method of measuring the speed of light was to direct a light source towards a rotating mirror, have the rotating mirror reflect the light to another mirror. The second mirror then reflects light back at the rotating mirror. The rotating mirror then reflects that light back towards the source.

Foucault's Rotating Mirror Experiment

1. Light is sent at the rotating mirror (A).

2. Light is reflected from rotating mirror (A) to stationary mirror (B).

3. Light is reflected from stationary mirror (A) to rotating mirror (B).

4. Light is reflected from rotating mirror towards source.

5. Using difference of the angles of light waves (1) and (4), the speed of the mirror's rotation, and the distance light travels, the speed of light is calculated.

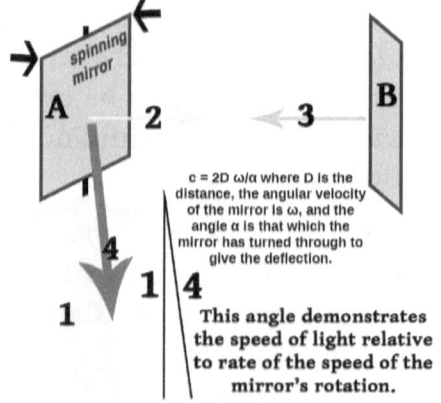

$c = 2D \, \omega/\alpha$ where D is the distance, the angular velocity of the mirror is ω, and the angle α is that which the mirror has turned through to give the deflection.

This angle demonstrates the speed of light relative to rate of the speed of the mirror's rotation.

In 1863, Gregor Johann Mendel (1822 – 1884) finished a series of experiments with pea plants which established many of the laws of what would later be known as *Mendelian inheritance*. Mendel is known as the *Father of Modern Genetics*. Mendel also experimented with breeding and hybridization of plants.

Mendel, born in Austria (now Chech Republic) and lived Saint Thomas' Abbey as an Augustinian friar. Mendel initially worked with mice, but his clergy did was opposed to him studying animal sex, so Mendel switched to plants.

The vast majority of Mendel's published works, however, have nothing to do with genetics or plant and animals at all, but astronomy and meteorology instead.

In 1865, Mendel founded the *Austrian Meteorological Society.*

Staring at the Sun

Solar physics have always been a hotbed of paradoxes. The ancient Egyptians worshiped the Sun as a falcon-god's eyeball. To the Greeks it was a chariot-wheel rolling around. During the industrial revolution it was a burning ball of coal, but that would have burned out in a couple hundred million years.

Enter the Kelvin–Helmholtz mechanism – heating by compression.

William Thomson, 1st Baron Kelvin (1827 – 1907) was a Scots-Irish mathematical physicist and engineer. Kelvin is best known for the absolute-zero scale of temperature named in his honor. William (Lord) Kelvin accurately predicted the value of absolute zero as (-)273.15 Celsius/(-)459.67 Fahrenheit.

> *Whoever, in the pursuit of science, seeks after immediate practical utility may rest assured that he seeks in vain.*

> Hermann Helmholtz
>
> Academic Discourse (Heidelberg 1862)

William Kelvin and Hermann Ludwig Ferdinand von Helmholtz (German 1821 – 1894) proposed that the Sun was powered by heat 'squeezed out of the constituent gases' by the force of gravity.

In this model, the Sun was not actually generating any heat, but was simply exhausting what it already had and, according to Kelvin, "the sun is actually cooled from year to year."

The Kelvin–Helmholtz mechanism was hotly contested and fairly paradoxical. If, for instance, heat became repelled from a gravitational center, then the inner parts of the Sun must be very cold – a problem that would persist when Eddington employed it later.

Another problem is that the heat-squeeze mechanism wouldn't last very long. The Kelvin–Helmholtz mechanism would only light the Sun for about 8.9 million years at the present rate of discharge. William Kelvin argued that the Earth was only millions, not billions of years old.

In 1862 Kelvin published *On the Age of the Sun's Heat* in which he stated "the sun has not illuminated the earth for 100,000,000 years" saying, instead, that the Sun has existed maybe 'ten or twenty million' years and, as a luminary object, "has not done so for 500,000,000 years."

Lord Kelvin went on to calculate that the Earth, from a molten state, would cool down to a level animals could walk on in exactly 98 million years, further establishing Cavendish Earth theory. Lord Kelvin's calculations demonstrate that any heat from one billion years ago would certainly be long gone.

Why so young? Because volcanoes are erupting. According to the Kelvin-Helmholtz mechanism all there was was latent heat from a hot time when Earth was forged. Within that theory, there is no significant source of heat.

If nothing is replenishing heat in center of the Earth, then when magma reaches the surface and solidifies, it will never be magma again. That heat has been lost to Earth forever. By observing the amount of volcanic activity on Earth and correlating that with loss of heat, Kelvin arrived at the conclusion that the whole Earth was a molten mass merely 98 million years ago.

Lord Kelvin was greatly offended at Darwin when Darwin had proposed that Earth was more than 300 million years old. Kelvin even went after Darwin directly, openly criticizing his assessment of the age of the Wealds of England. Those could not exist if, at the same time, the Earth were molten.

> "What then are we to think of such geological estimates as 300,000,000 years for the "denudation of the Weald"? ... with possibly Channel tides of extreme violence, should encroach on a chalk cliff 1,000 times more rapidly than Mr. Darwin's estimate of one inch per century?"

Helmholtz demonstrated that the 'immediate practical utility' of erosion was not a comprehensive theory. Floods, tides, sinkholes, ice ages, and differing rates of precipitation would each alter the rate of denudation. The example that Darwin gave was not sufficiently comprehensive to draw broad-based assumptions. The denudation of the Wealds is obviously insufficient as a means of gauging Earth's age

Helmholtz's own mechanism – a Sun powered by squeezing – was also an 'immediate practical utility' and was ultimately a failure.

The world of science now hinged on the assumptions of Copernicun gravity, flying photons, Cavendish Earth, gravity equality, 6.67, strange gases, random evolution, Darwin's timeline, neutral gravity, and that 'immediate practical utilities' justify momentarily heating of large bodies of gas.

In August of 1864 August Wilhelm von Hofmann (1818 – 1892) became professor of chemistry and director of the chemical laboratory at the University of Berlin. Hofmann was brilliant German who attended the University of Giessen, specialized in organic chemist, and mentored many great chemists who came later.

James Clerk Maxwell

James Clerk Maxwell (1831-1879) was a Scottish master of wave physics in the field of *mathematical physics*. In 1865, Maxwell published *A Dynamical Theory of the Electromagnetic Field* in which he demonstrated wave physics and predicted that light was an electromagnetic transmission.

Maxwell was thoroughly home educated by a loving Victorian era mom and began attending public schools at age 10. Maxwell's academy education went on until he was 16.

Maxwell effectively defined light and light interactions, merging the fields of electricity and light in a cohesive *electromagnetic* theory. This comprehensive set of electromagnetic concepts are called *Maxwell's Equations*.

In 1861 Sir William Crookes (1832 – 1919) was doing spectroscopic research and discovered an element with a very bright green spectral emission line, so he named it after the Greek work *thallos* which means 'green shoot' – *thallium*.

Crooks was the oldest child of a family with 16 children. Crooks attended the Royal College of Chemistry (1845 – 1872) where he assisted Wilhelm von Hofmann. That association allowed him to join the Royal Society.

Alfred Nobel

In 1867, Alfred Bernhard Nobel (1833 – 1896), a Swedish inventor, scientist, researcher, and businessman invented dynamite by mixing *nitroglycerin* with *diatomaceous earth (DE)*. The DE stabilized the nitroglycerin allowing it to be safely handled. Unlike most forms of gunpowder, dynamite would only detonate if you really tried to ignite it – if you lit a fuse or ignited it electrically.

Dynamite was one of 355 patents held by the Alfred Nobel.

In 1871 Crookes published *Select Methods in Chemical Analysis,* advancing on the spectroscopic methods of Kirchhoff and Bunsen.

Johannes Diderik van der Waals (1837 – 1923) was a Dutch (Netherlands) carpenter's son with a poor man's primary education. In his thesis in 1873 titled O*n the Continuity of the Gaseous and Liquid State*, van der Waals posited concepts of molecular attraction and molecular volume.

This new concept was contrary to the *Ideal Gas Law*. According the ideal gas law, atoms and molecules do not attract, but only repel. Van der Waals became a great theoretical physicist, thermodynanamist, and molecular scientist.

In 1874 Cambridge University opened the *Cavendish Laboratory* in honor of Henry Cavendish. Cavendish laboratory is located in Cambridge, Cambridgeshire, England.

In 1875 Crookes invented the *Crookes tube.* The Crookes tube is a vacuum tube used to experiment with electrical discharges. A vacuum tube consists of a region of space which has been vacated of matter.

When running normally, the Crookes tube projects a Maltese cross onto the center of the end of the tube. If a magnetic field is applied to the situation, the shadow of the cross becomes displaced, showing that light or electron discharges are influenced by magnetic charges.

The Crookes tube later led to cathode tubes where the anode and the cathode are at opposite ends.

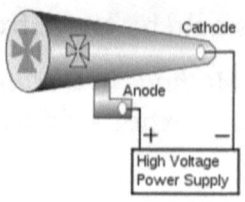
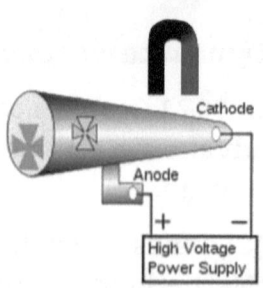

Meanwhile, in the Netherlands, Hendrik Lorentz (1853-1928) received his doctorate at the University of Leiden in 1875 for a thesis titled *On the Theory of Reflection and Refraction of Light*. That thesis was a study of, and advancement of, the electromagnetic models of James Clerk Maxwell.

During the University of Leiden Inauguration on 25 January 1878, Lorentz delivered a famous lecture on *The Molecular Theories in Physics* – heralded as the beginning of the age of atomic chemistry.

Lorentz theorized that atoms consist of charged particles. He also suggested that the <u>oscillations of these charged particles caused light</u>. This observation was later supported with the observation that when an electron charge loses energy, and the electron 'falls to a lower orbit' so to speak, that energy is emitted as light.

Lorentz developed Lorentz aether theory – a theory with subtle difference to Michelson-Morley's Luminiferous Aether theory.

Lorentz refined and advanced the electromagnetic models of James Clerk Maxwell and extended those physical principles to encompass atomic theory.

In 1877 Max Karl Ernst Ludwig Planck (1858 – 1947), a German theoretical physicist, left the University of Munich to study at Friedrich Wilhelm (Berlin) University (1809 – 1945) in Berlin. There, he studied under Hermann von Helmholtz and Gustav Kirchhoff.

Planck wrote that Helmholtz was never quite prepared, spoke slowly, bored his listeners, and miscalculated endlessly. Plank wrote that Kirchhoff delivered carefully prepared lectures which were dry and monotonous. Planck became close friends with Helmholtz.

Heinrich Rudolf Hertz (1857-1894) was a German physicist who confirmed the existence of electromagnetic radiation through radio waves, confirming the predictions and observations of James Clerk Maxwell.

Hertz studied under Gustav R. Kirchhoff and Hermann von Helmholtz, obtaining his PhD from the Friedrich Wilhelm (Berlin) University in 1880 and performing post-doctoral study under Helmholtz until 1883.

The measurement of the frequency of a radiation is named after Heinrich Hertz. One Hertz (Hz) equals one cycle per second. Sound and alternating current electricity are typically expressed in Hertz (Hz). In America, household electrical alternating current (AC) power is typically operated at 60Hz or 60 cycles per second. The range of human hearing is 20Hz up to 20,000 Hz.

Shaking Things Up

Earthquakes have occurred throughout all of human history. Scientifically, measurement of earthquake effects are documented back to 132 CE, when Zhang Heng of China's Han dynasty invented *Houfeng Didong Yi* – an "instrument for measuring the seasonal winds and the movements of the Earth".

Jean de Hautefeuille (1647 – 1724) was a French abbé, physicist and inventor who invented a seismograph. Jean also proposed the mechanism for a piston-driven internal combustion engine, although his theoretical version used gun-powder for fuel instead of gasoline. He also worked Ewing, Gray, and Milne.

Sir James Alfred Ewing (1855 – 1935) was a Scottish physicist and engineer famous for inventing the modern horizontal seismograph. He is also well-known for his works on the magnetic properties of metals.

Thomas Lomar Gray (1850 – 1908) was a Scottish engineer best known for his work in seismology

John Milne (1850 – 1915) was a British and mining engineer who also worked on the horizontal seismograph.

Thomas Corwin Mendenhall (1841 – 1924) was an American autodidact physicist and meteorologist most famous for his invention of the gravimeter.

The talents of Ewing, Gray, Milne, and Mendenhall were combined in Japan – an earthquake rich region on the eastern Pacific rim – where great achievements occurred.

In 1880 Milne invented the horizontal pendulum seismograph. This design – the standard for most modern seismograph models – could not only detect earthquakes, it could detect different types of waves, and it could estimate the velocities of those waves.

This device greatly differed from standard *vertical seismographs* in that the horizontal seismograph measured horizontal motion. Earthquakes at that time were generally thought to be up-and-down displacement. Milne's horizontal seismograph demonstrates that earthquakes involve horizontal displacement.

These findings preceded the development of *plate-tectonic theory* and would later be used as evidence supporting that theory.

In 1880 Max Planck completed his doctorate with a thesis titled *Equilibrium states of isotropic bodies at different temperatures.* Plank redefined thermal laws using Planck scale size, and created predictions of maximum and minimum temperature with that.

Following the invention of dynamite, Alfred had adjusted his business focus to the production of dynamite and the forging of cannons. Dynamite and the related principle of *controlled detonation* multiplied the use and effectiveness of explosives in both war and peace.

One French newspaper characteristically referred to "Dr. Alfred Nobel, who became rich by finding ways to kill more people faster than ever before..." This was an attitude typical among the French elite.

The French protest against the Alfred Nobel's technology became most apparent when France endeavored to build the Panama Canal (1881 – 1894). The French approach to that project was *very* labor-intensive because the French refused to use dynamite.

If the French had used dynamite on the project, Alfred Nobel would have become the richest man in the world.

In 1887 Hertz published *On Electromagnetic Effects Produced by Electrical Disturbances in Insulators* which demonstrated, among other things, that electromagnetic waves travel in a vacuum without air.

Hertz confirmed that light travels as waves within a medium and that that medium was not air.

He also made fundamental predictions as to how wavelengths corresponds to the amount of energy transmitted – longer waves have less energy – principles later codified as laws of *thermal radiation.*

Since the experiments of Hertz used what may be described as the first combination of emission and reception of wireless electromagnetic radiation, Hertz is sometimes credited with inventing or discovering radio waves.

Michelson-Morley

Around 1869, Albert Abraham Michelson (1852-1931), a Polish-born Jewish American officer in the United States Navy, had began planning to repeat the experiments of Léon Foucault.

Aristotle (384 BC – 322 BC) wrote, 'Light is due to the presence of something, but it is not a movement.'

The speed of light had been reasonably established, but established relative to what, Michelson pondered.

Albert Michelson, who was raised in the mining towns of Murphy's Camp, California, and Virginia City, Nevada, worked on these questions with Edward Morley's assistance for several years.

Edward Williams Morley (1838-1932) was born in Newark, New Jersey. A sickly child, Ed was home-schooled by his father until he was nineteen years old. Morley became well-known for his work in optics and astronomy and for his work with Albert Michelson.

Both men are both best known for the Michelson-Morley Experiment of 1887.

Team Michelson-Morley postulated that light was made of waves – waves or disturbances within a material medium. This meant that light did not consist of flying photons. Very differently than flying photon theory, wave-model light-wave transmission does not involve net displacement of matter or particles or subatomic particles.

Michelson's concept of light-wave transmission was that light-waves and sound-waves were similar in principle.

Sound energy consists of waves transmitted across a medium. Sound consists of vibrations transmitted across the air.

Sound transmission is not the *moving* of air.

As sound is transmitted, individual air atoms offset away from the source of sound and then back towards the source of sound. After sound is transmitted across air, the air is found in its original location. This a longitudinal wave. The medium itself is moving - compressing and decompressing, and transmitting sound at 747 kilometers per hour.

Since this is a vibration and the medium (air) remains (effectively) unmoved, sound is transmitted at equal speed in every direction simultaneously – that's 747 kilometers per hour up, down, right, left, front, and back.

Not only that, but an atom may be involved in transmitting multiple sounds in multiple directions simultaneously. Sound does not crash into other sound. Sound is not a wind blowing 747 kilometers per hour.

Vibration of air is sound. When air moves, that is not sound. The air collectively moves, that is wind.

In light-wave theory, light-waves are transmitted similarly to sound waves, but they are lateral waves instead of longitudinal waves.

A lateral/electromagnetic wave does not physically displace the medium. A lateral wave biases the medium instead.

Suppose we transmit a sound wave (longitudinal) and a light wave (lateral) across a medium of hydrogen. Watching a group of hydrogen atoms conduct sound would show hydrogen transmitting sound waves as alternations of compression and dilation.

To watch hydrogen conduct light, we have look much more closely. When a light wave arrives, a wave of photons occurs concurrent with the bias of the field - either positive (+) or negative (-). When a positive (+) wave is passing, positive (+) photons occur and few electrons occur. When a negative (-) wave is passing, negative (-) photons occur along with extra electrons.

The medium remains static and the speed of light is 299,792,458 meters per second in every direction at the same time.

In flying photon theory, light is transmitted like wind. Light is accomplished by flying photons and <u>there is no medium</u>. Within flying photon theory, light consists of winds blowing 299,792,458 meters per second in every direction simultaneously.

Modeling light waves similarly to sound waves, Michelson-Morley determined that if you split a wave in half and then recombined that energy at opposition – where half of the waveform in inverted from the other half – then that energy should cancel itself out.

They proved this principle to be true with the invention, construction, and demonstration of the *interferometer*.

An interferometer works by splitting a wave into two equal halves, sending them down tubes of slightly different length, then recombining the waves where one of the waves is a half-cycle off. One of the waves has been delayed for one-half of one wavelength longer than the other half of the wave. When these opposite forces affect each other, their energy totals zero. The two halves energetically cancel each other out.

Interferometer

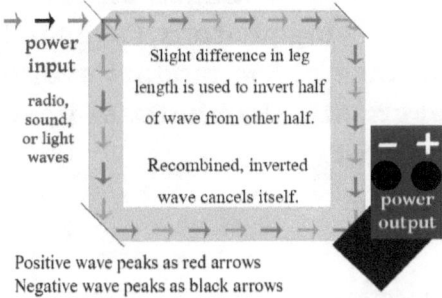

Positive wave peaks as red arrows
Negative wave peaks as black arrows

This graphic was not the form of Michelson-Morley's interferometer, but demonstrates the principle they observed.

Interferometer

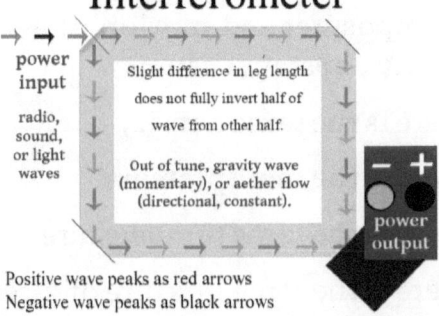

Positive wave peaks as red arrows
Negative wave peaks as black arrows

Through Michelson-Morley's experiments, it was established that light traveled at a constant speed relative to the surface of the Earth. If the speed of light was not constant relative to the surface of the Earth, the interferometer would not work. It would constantly go out of tune because the Earth keeps spinning.

The Quest for Aether Speed

After Michelson-Morley thoroughly demonstrated that the speed of light was constant relative to the surface of the Earth, they then proposed that the speed of light was constant throughout the universe.

That is a completely different concept.

The world of science now hinged on the assumptions of Copernicun gravity, flying photons, Cavendish Earth, gravity equality, 6.67, strange gases, random evolution, Darwin's timeline, neutral gravity, Helmholtz heat, and a <u>universally</u> constant speed of light.

What if the speed of sound was universally constant?

The speed of sound is determined by the density, pressure, temperature, composure, and motion of air. If the speed of sound is universally constant then;

All air everywhere is the same density

All air everywhere is the same pressure.

All air everywhere is the same temperature.

All air everywhere is the same mixture of gases.

There is no wind.

All air everywhere never moves.

If any of these statements are not true, then the speed of sound is not universally constant.

Actually, Michelson-Morley never did propose that the speed of light was constant relative to the Earth. That implication of their experiments – <u>the most important one</u> – was something they claimed was not true.

They found it to be true, maybe they knew that it was true, but they went on trying to prove that it was not true.

To make that not true, and to accommodate a universally

constant speed of light, Michelson-Morley designed their ae-ther – the Luminiferous Aether (LE)– as being universally static. That way, light traveling through the LE – conducted by the LE as waves – would be conducted at the same speed everywhere across the universe.

Michelson-Morley or their counterparts were perhaps a little confused as to the *nature* of light waves. If light waves were like sound waves, than those are compression or *longitudinal waves* later called *gravity waves*. For the aether to transmit lon-gitudinal waves at the velocity of light, it would have to be very dense. Solid.

Some physicists of the time joked that the "Luminiferous Ae-ther of Michelson-Morley must have the tenacity of steel" to accomplish light-speed transmissions.

Some people were confused at the idea of Earth (having the tenacity of dirt) plowing through an aether of steel, and how, on the surface of Earth, you wouldn't feel so much as a breeze.

In the *illustration* below, where Aether Flow is represented as 5,000 meters per second, the speed of light would be different in every direction. That is what Michelson-Morley were look-ing for.

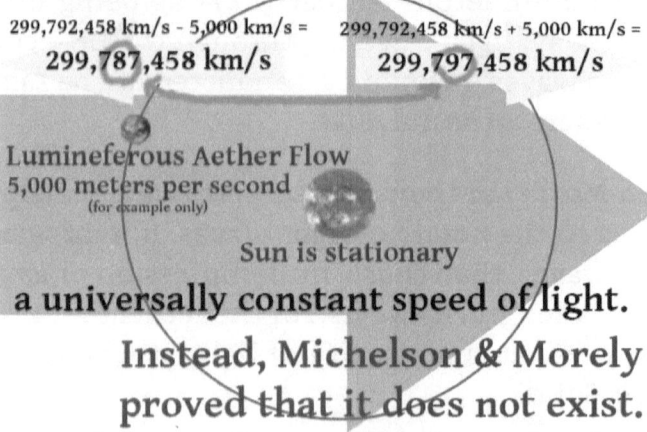

Michelson & Morely were looking for

299,792,458 km/s - 5,000 km/s =
299,787,458 km/s

299,792,458 km/s + 5,000 km/s =
299,797,458 km/s

Lumineferous Aether Flow
5,000 meters per second
(for example only)

Sun is stationary

a universally constant speed of light.

Instead, Michelson & Morely proved that it does not exist.

Result: Lumineferous Aether Flow = 0 (zero)

Since Earth spins and wobbles and orbits the Sun, the Earth had to be moving *through* this LE. The Sun has movement relative to the galaxy, so the Sun had to be plowing through it, too. And the Milky Way galaxy is moving relative to other galaxies, so our whole galaxy had to be plowing through it.

While plowing through the Luminiferous Aether, the difference between the speed of an object and the speed of the LE was termed 'aether drift, drag, or flow'. Since the speed of light was supposed to be constant relative to the LE instead of constant to individual objects, the aether flow was something that experiments could detect and measure.

That could be done because if the speed of light were constant relative to a LE, then the speed of light *must* be different in virtually every direction on Earth.

Michelson-Morely Experiment

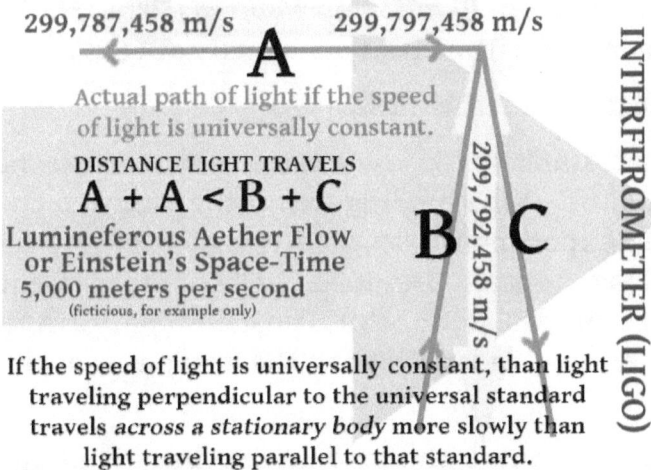

299,787,458 m/s   299,797,458 m/s

A

Actual path of light if the speed
of light is universally constant.

DISTANCE LIGHT TRAVELS
$$A + A < B + C$$

Lumineferous Aether Flow
or Einstein's Space-Time
5,000 meters per second
(ficticious, for example only)

B **C**

299,792,458 m/s

INTERFEROMETER (LIGO)

If the speed of light is universally constant, than light
traveling perpendicular to the universal standard
travels *across a stationary body* more slowly than
light traveling parallel to that standard.

In fact, that is exactly how Michelson-Morley said that they could prove that their Luminiferous Aether existed – they would prove that their LE existed by proving that the speed of light was universally constant. They would prove the speed of light to be universally constant by proving that the speed of light was <u>not</u> constant relative to the surface of the Earth.

They would do that by *precisely* measuring the speed of light in opposing directions with an interferometer.

In the graphic above, it is shown that if there is an aether flow, that the distance light travels perpendicular to that flow would be a longer path than the light traveling parallel to that flow. If the speed of light is constant in every direction relative to an immobile LE, an interferometer such as LIGO (Laser Interferometer Gravitational wave Observatory) would detect this delay.

So Michelson-Morley built their latest and greatest interferometer that would show how fast the Earth is moving relative to the universal speed of light and announced their intentions to the world.

The famous Michelson-Morley experiment was scheduled to

occur in 1887 – exactly two hundred years after Isaac Newton's publication of *Philosophiæ Naturalis Principia Mathematica*. It began in April and lasted until July of 1887.

That detected no aether flow.

When that didn't work, the experiment was extended into October. Most of the funding was suspended and the second leg of the experiment was quietly executed my Michelson alone. This extension provided time to improve equipment and also to let the Earth's orbital position change in case Earth was 'aether-flow-neutral' during the first phase of the experiment.

That didn't improve things any. The improved equipment showed better than ever that the speed of light was constant relative to the surface of the Earth and <u>not</u> to any universal standard.

Because Michelson-Morley's aether model (Luminiferous Aether (LE)) was built around a universally constant speed of light, when their experiment failed, their LE model was greatly discredited.

The concept of a universally constant speed of light was not discredited.

Michelson worked in the primarily in the private sector after that.

In 1887 Alfred Nobel invented *ballistite* - a mixture of 40 percent nitrocellulose and 60 percent nitroglycerin. Cut into flakes, this explosive was a superior, near-smokeless artillery propellant and became a standard munition for the next 75 years.

In 1888, a French newspaper misunderstood and misreported the death of Alfred Nobel's brother as the death of Alfred instead. The obituary which the living Alfred Nobel read was titled *The Merchant of Death is Dead.*

Lorentz, meanwhile, postulated the concept of 'local time', suggesting that this would simplify understanding Michelson-Morley's findings and the Luminiferous Aether. Although Lorentz theorized 'local time', this 'local time' would be connected to the 'universal clock' via the universal speed of light.

But that proposal did not make Michelson-Morley's findings into a constant speed of light. Lorentz found a different workaround a few years later.

In 1889, Arthur Schuster published two letters in the British magazine *Nature* in which he posited the concept *anti-atoms* and whole *anti-matter solar systems* which, if one came in contact with a ordinary solar system, would annihilate both solar systems, resulting in energetic output.

In 1889 Alfred Nobel signed a deal selling 300 tons of ballistite to the Italian government and then sold his Italian ballistite patent to the Italian government.

In France, Nobel was accursed of committing 'crimes of high treason' and 'espionage' for arming Italy – part of the Triple Alliance and an unofficial enemy of France. France withdrew Nobel's license to conduct business in France and confiscated property from Nobel's factories and laboratories.

In 1891, Alfred Nobel moved to Italy, where his estate – Villa Nobel – is now a museum in Sanremo, Italy.

Lorentz Compression

In 1892, Lorentz found a method to describe how that the speed of light could be universally constant, the Luminiferous Aether could exist, that LE could be universally static, LE could cause aether drift, <u>and</u> that that LE flow could never be physically detected. All at once.

Now known as *Fitzgerald-Lorentz compression*, Lorentz compression dictates that objects which are in motion relative to the universal standard of light compress with respect to the difference in motion between the object and the LE.

In other words, Fitzgerald-Lorentz compression means 'compression proportional to movement relative to the standard of the Luminiferous Aether.'

Why? Because that is the _only way_ that a universal speed of light *could* exist given the evidences of Michelson-Morley, Nikola Tesla, James Clerk Maxwell, and many others.

Lorentz Compression

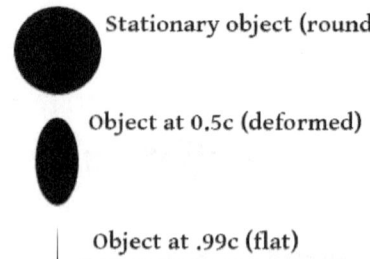

Stationary object (round)

Object at 0.5c (deformed)

Object at .99c (flat)

If the speed of light (c) is universally constant and the speed of light is constant as measured on a moving object, then moving objects must contract proportional to motion.

If the speed of light is locally constant and not universally constant, than Lorentz transformation need not exist.

James O Harris

The faster you go the shrinker you get.

Lorentz transformation not only violates common sense, it violates basic physics as well. Why would matter compress when moved? What is causing this compression?

Lorentz succeeded in bending the shape of the universe to the will of the speed of light.

The Age of Paradoxes had begun.

The world of science now hinged on the assumptions of

Copernicun gravity, flying photons, Cavendish Earth, gravity equality, 6.67, strange gases, random evolution, Darwin's timeline, neutral gravity, Helmholtz heat, constant light speed, and that every object in the universe bends and stretches to keep the speed of light constant.

Black body Dynamics

Wilhelm Carl Werner Otto Fritz Franz Wien (Wilhelm Wien) (1864 – 1928) was a German physicist who, in 1893, deduced *Wien's Displacement Law*.

Wien's displacement law is the law of black body radiation. Black-body radiation has to do with the relationship of a wavelength and a temperature. The higher the temperature of an object, the shorter the wavelength of its emissions – an reality proposed by Heinrich Hertz six years earlier.

When speaking of stars, the bluer they are (the shorter the wavelength), the hotter they are. The redder they are (the longer the wavelength) the colder they are.

Wein's Displacement Law dictates that heat in an body of even temperature radiates equally in every direction without regard for the influence of gravity.

In 1895 William Crookes identified the first known sample of *helium*.

On 27 November 1895, Alfred Nobel (1833 - 1896) signed his last will and testament, giving the largest share of his fortune to a series of prizes in Physics, Chemistry, Physiology or Medicine, Literature, and Peace – the Nobel Prizes.

Although these prizes were to begin being awarded immediately, distribution did not begin until five years later.

In 1896, Henri Becquerel (1852 – 1908) discovered the existence of *cosmic rays.* Cosmic rays occur in Earth's upper atmosphere and are denoted by cascading discharges of light.

Cosmic rays are thought to occur from primarily outside of our Solar System and from beyond our galaxy.

Nikola Tesla posited around that time that cosmic rays were source power for gravity.

In 1897 Max Planck published *Treatise on Thermodynamics* which included proposed a thermodynamic basis for Svante Arrhenius's theory of electrolytic dissociation.

Cathode Rays and Electrons

In 1897, Joseph John Thomson (1856 – 1940) proposed that cathode rays were streams of subatomic particles. He called them *corpuscles* the same way that Newton referred to photons. They would later become known as *electrons*.

A cathode tube consists of a vacuum tube with a cathode at one end and an anode at the other. The cathode is the electrode where negative (-) direct current (DC) electricity is applied. The anode is the electrode where positive (+) direct current electricity is applied. When energized, a cathode tube produces light similar to the way a neon light produces light. Neon lights use a tube filled with neon gas, but cathode rays use vacuum tubes with nothing inside.

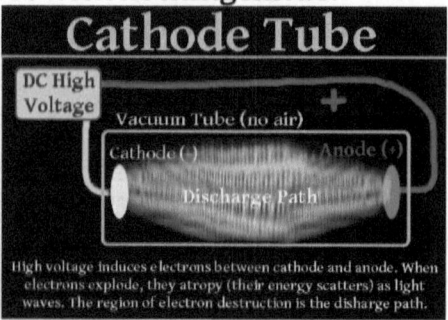

J.J. Thomson was adept at science since a young age, and was accepted to Owens College (now University of Manchester) at the age of 14. Thomson begun attending Trinity College at Cambridge in 1876.

On December 22, 1884 J.J. Thomson had became Cavendish Professor of Experimental Physics of Cavendish Laboratory at the University of Cambridge. Zt this point in time, is was already known that when high voltage is applied to a cathode tube, the tube lights.

Thomson placed a box over the cathode and put a slit in it. When he did this, the discharge path changed.

Thomson suggested that the discharge path consisted of 'flying electrons' which were returning to ground. The discharge path became known as *cathode rays,* shown in in the illustration.

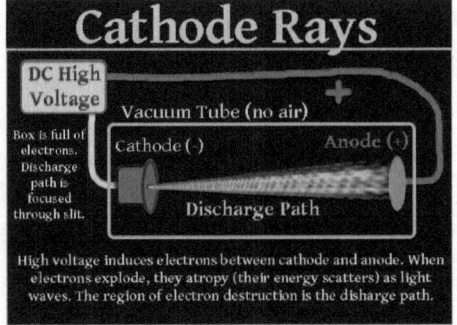

Thomson showed that when an external field is applied to the cathode rays, that the discharge path changes. Thomson reasoned that the external field applied attractive magnetic force to the electrons, thereby altering the physical trajectories of flying electrons. This, Thomson reasoned, was an exchange of momentum. This being an exchange of momentum, Thomson proposed this as evidence that electrons have mass. Thomson calculated that flying electrons had very low mass and very high charge characteristics.

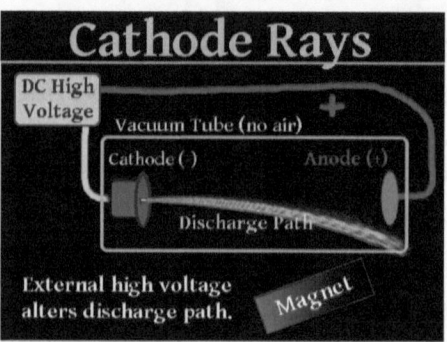

The strange thing is that when an external magnetic field is applied, the 'flying electrons' – as observed – do not return to ground. If that discharge path is a stream of electrons in this illustration, where are those electrons going? Those electrons are not returning to their source.

Where are electrons going? What is satisfying the difference of potential? Power is being consumed, right? This is not a free energy/free lighting device with unlimited output. No. Energy is being consumed. But the electrons (if that discharge path is flying electrons) are <u>not returning to ground.</u>

And how is it that you are *seeing* an electron? If the electron is discharging light, then you are not seeing electrons. You are seeing photons. Electrons are degrading into photons. If an electron is discharging light, it is dissipating. If the electron dissipates, than the electron does not travel, or travels very little, before being destroyed.

The fact that light is emitted perpendicular to the discharge path demonstrates that <u>the discharge path is where grounding occurs.</u> The difference of electrical potential is being destroyed <u>in the discharge path.</u> The energy of that discharge is being entropied as the light which the observer witnesses.

This is a field reaction.

What is the field?

And for that matter, where did the electrons come from?

If atoms have only a certain, limited number of electrons, where are the trillions of atoms that are missing their electrons – the electrons inside of that cathode tube?

Did they come from a copper coil inside of a generator? No. What would happen if you took all the electrons out of section of copper coil? The answer is that all chemical bonds would dissipate. Without any negative charge off-setting the positive atomic nuclear repulsion, the copper would burst into a gaseous state.

What about the magnets? What would happen if you took all of the electrons out of a magnet? Then you would not have a magnet anymore. The whole mass would be positively charged and, like copper, it would basically explode.

If the magnet did not explode, then it would rearrange to it's least resistive natural state: Not a magnet.

The fact is that electrons cannot be 'separated' from atomic nuclei because atomic nuclei *induce* (cause, form, create) electrons. Electrical generators may be thought of as very large virtual atoms which pump out electrons.

Before Thomson's cathode ray experiments, atoms were thought to consist of smaller atoms with hydrogen being the smallest unit. Thomson's proposition that electrons were atomic pieces opened up a whole new field of investigation: Subatomic particles.

Thomson's experiment does, in fact, demonstrate the existence of electrons *and photons*. The experiment does not, however, demonstrate that electrons have any mass.

Also note that the cathode rays are <u>not</u> stronger nearer the source. If cathode rays consisted of electrons flying out of the cathode slit, then discharges should certainly be stronger there, and diminish with distance. That is not the case.

Furthermore, there is a 'dark space' between the cathode and the discharge path. If electrons were traveling, they would have to cross the dark space. But electrons are not evidenced as existing there.

In fact, there is not just one dark space, but four independent, different dark spaces. Beyond the surface of the *cathode* is the *Aston dark space,* then there is a layer called the *cathode glow,* then there is a layer called *Crookes dark space* followed by a region called the *negative glow.* Then there is the region called the *Faraday dark space*, followed by a column called positive glow. Then there is the *anode dark space* before the *anode glow* and the *anode.*

That's *four different* dark spaces where electrons are not traversing or occurring or exploding.

If the electrons are not traveling out of the cathode, where are they coming from? They are occurring *between* the cathode and the anode. The difference of potential between the cathode and anode induces electrons.

Electrons are temporary, transient particles which occur and then dissipate. They behave the same way photons do.

Contrary to the observations above, Thomson's electrons were canonized as permanent, bosonic, traveling particles having mass.

The world of science now hinged on the assumptions of Copernicun gravity, flying photons, Cavendish Earth, gravity equality, 6.67, strange gases, random evolution, Darwin's timeline, neutral gravity, Helmholtz heat, constant light speed, Lorentz compression, and massive, flying electrons.

When Thomson placed the box and the slit on the anode instead of the cathode, he found *canal rays.* These, he supposed, were positive counterparts of electrons. Once again the discharge path of canal rays did not indicate any movement of particles.

Later, in the field of quantum mechanics, Paul Dirac would formulate a concept of electrons which predicted a counterpart to electrons – positrons.

This concept is also contradicted by cathode/canal ray experiments, because the discharge paths of *canal rays* also contain electrons – not positrons.

In 1894 Max Planck was hired by electrical companies to build a better light bulb. In an effort to improve light bulb technology, Planck was tasked with figuring out how heat moves. *Wien's Law,* came up short when measuring the radiation and speed of low-frequency transmissions of energy.

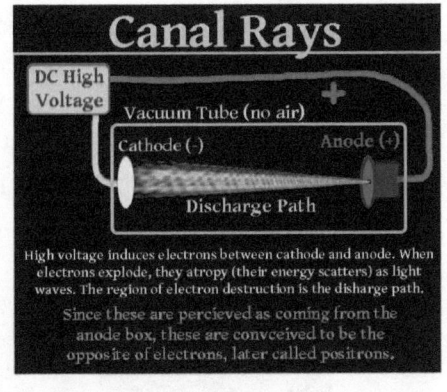

Planck tried solving the problem using uncertainty with a "principle of elementary disorder". That concept floundered experimentally.

Grudgingly, Planck then employed the Ludwig Botzmann (1844 – 1906) statistical model in "an act of despair ... I was ready to sacrifice any of my previous convictions about physics.". That sacrifice would win Planck a share of a Nobel Prize in Physics (1933) "for the discovery of new productive forms of atomic theory."

The Man Who Wired the World

In 1896 Nikola Tesla and George Westinghouse, who had beat out the Edison Company for a government contract, completed the first major alternating electricity power station and the first hydro-electric power plant at Niagara Falls,

Buffalo, New York, Power House Number 3.

Nikola Tesla (1856-1943) was a Serbian-born American inventor, electrical engineer, and futurist. His inventions included alternating current generators, alternating current motors, alternating current electrical transmission, alternating current transformation, polyphase current technologies, radio-wave emitter and receiver technologies, and obscure inventions including technology equivalent to a PET scan, zero-point energy transformation technology, longitudinal wave technology, electromagnetic heating and cooling technologies, remote control guidance systems, longitudinal weapon technology, and gravitational manipulation technology.

The outstanding success of Nikola Tesla's AC technology changed the face of the planet. All major power grids now operate on Tesla's *Alternating Current* technologies.

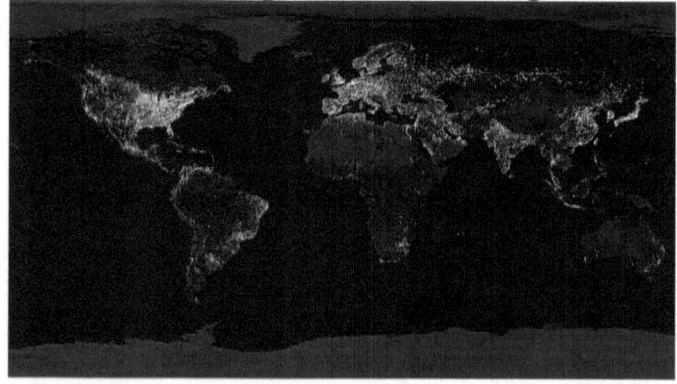

The many inventions and advancements made by Nikola Tesla were not academic in nature and Tesla's laboratories were arsoned on two different occasions, resulting in untold losses of written documentation.

Tesla held that space consisted of an aether and/or an *omni*. Tesla opposed Einstein's theory of 'nothing-space' space-time. Einstein and Tesla were friends who regularly corresponded with each other.

Nikola Tesla was a master scientist and genius inventor. His contributions, however, had little direct effect on modern scientific theory.

"The evolution of electric power from the discovery of Faraday in 1831 to the initial great installation of the Tesla polyphase system in 1896 is undoubtedly the most tremendous event in all engineering history"

-Dr. Charles F. Scott, Professor Emeritus of Electrical Engineering at Yale University and former President of the American Institute of Electrical Engineers

In 1898 Pierre (1859-1906) and Marie Curie (1867-1934) published their discovery of polonium and radium. Their working history included discovering the thermal limits of magnetic fields and magnetic substances, now known as the *Curie Limit*. Marie Curie was the person who coined the term 'radioactive'.

In 1899 Max Planck introduced the Planck constant – creating a new concept of 'the smallest size of anything'. The 'smallest thing' had once been atoms, then atomic nuclei, then subatomic particles, then energy which does not take up any space.

At some point in this scale, the laws that govern atoms just don't make sense anymore. That is presently considered to be the Planck size.

The Planck constant is the *Grinch Limit*. Remember Whoville, a city on a snowflake? At Planck sizes, there is said to be no who, what, why, when, or Whoville.

Is the Grinch Limit fact, fiction, or mathematical naval-staring? Is there a reason that physics would cease to be physics on very small scales?

If space is a naked, vacant void then: Yes.

If space is nothing than physics must terminate at some small scale.

If there is no material aspect of space, then there is nothing between matter and nothing to conduct communication between units of matter.

As Isaac Newton observed, if space is a vacant void of nothing, then *our* laws of physics need not apply on the other side of a vacant expanse, nor, perhaps, on the very next planet.

1900 - 1950

Many great men of history and lore have claimed to have 'seen the light'. On December 14, 1900, Max Planck quantized it. We can now mathematically calculate the enlightenment of various individuals. Max Planck quantized quantum mechanics with the following simple equation.

$$E = h v$$

Energy = Planck's constant * Frequency

Planck's constant is 6.626E-34 joule/seconds.

This is Max Planck's version of Wein's Law using the frequency instead of the wavelength.

Energy = speed of light / wavelength

What Planck's constant did that Wein's law did not do was establish a minimum value.

The value of Planck's constant was not figured out for many years. In fact, even Planck doubted the validity of this notion.

What makes this concept radical is, with Planck's equation, energy can now be defined in discreet units. While that's nothing new in the conventional sense of watts and volts, in the subatomic world this was completely new idea. This idea entertained the existence of a minimum threshold of energy – a minimum unit of energy.

This quantization would later seem quite natural in a world where everything is made of electrons or, later, photons, but Planck's constant would ultimately be determined to be hundreds of times smaller than a photon, allowing for energies exponentially smaller than a photon to exist. Those observations suggest the existence of objects much smaller than photons.

Planck became a powerhouse of academic influence and was later a key figure in bringing Albert Einstein's theories to the attention of Germany.

Prizes for the Greatest Benefit to Mankind

In 1901, the Nobel estate concluded its battle with the French government over a very hefty tax which the government sought from Alfred Nobel's estate.

Essentially, the French government wanted Alfred Nobel's profitable industries and explosive device patents to be owned by France.

When political pressure didn't work, the French government claimed that Nobel was a French citizen and tried to levy a heavy tax on his estate. When that didn't work, the French government tried to tax the estate as an export to Sweden. When that didn't work, they rehashed claims that he aided an enemy. When that didn't work, well, that just didn't work.

Alfred Nobel's legacy was returned to his homeland.

All that being settled, on December 10, 1901, the first Nobel Prize Ceremony was held in Stockholm, Sweden, exactly 5 years after the death of the great Alfred

1901

Nobel Prize in Physics

Wilhelm Conrad Röntgen

"in recognition of the extraordinary services he has rendered by the discovery of the remarkable rays subsequently named after him."

Nobel.

Nobel Prizes serve as hallmarks of scientific progress or achievement.

The Nobel Prizes center on what benefits mankind, so theoretical physics are rarely recognized. Theoretical physics have only theoretical value and are difficult to demonstrate as useful to mankind.

1901
Nobel Prize in Chemistry

Jacobus Henricus van 't Hoff

"in recognition of the extraordinary services he has rendered by the discovery of the laws of chemical dynamics and osmotic pressure in solutions."

Profits from Nobel's business holdings continue to fund the Nobel Prizes.

Among the first winners of Nobel Prizes was Marie Curie – the only woman to win two Nobel Prizes – 1903 and 1911 – and the only person to win Nobel Prizes in two different scientific fields – Physics and Chemistry.

1903
Nobel Prize in Physics

Pierre Curie, Marie Curie, and Henri Becquerel

"in recognition of the extraordinary services they have rendered by their joint researches on the radiation phenomena discovered by Professor Henri Becquerel"

Pierre and Marie Curie were responsible for many great advancements in science including the Curie Limit and the discovery of atomic radiation from radium.

Some people suggest that Pierre should have been recognized for the Nobel Prize awarded to Marie in 1911, but Pierre become ineligible.

Pierre Curie had been killed, run over by a horse-drawn wagon near the Pont Neuf in Paris on April 19, 1906. Only living people can get Nobel Prizes.

Ernest Rutherford

Ernest Rutherford (1871-1937) had shown up in England around 1900 and was building on the work of the Curies. In 1903 Rutherford and Frederick Soddy (1877-1956) published their "Law of Radioactive Change". This report was the first time that the eternal nature of atoms was called into question.

1904

Nobel Prize in Chemistry

Sir William Ramsay

"In recognition of his services in the discovery of the inert gaseous elements in air, and his determination of their place in the periodic system."

Rutherford and Soddy suggested that atoms could disintegrate as by radio-active decay, that that decay releases energy, and that atoms may be destroyed – converted into energy. Rutherford's concept of mass/energy proportionality was later codified by Albert Einstein as $e=mc^2$.

In 1904, Rutherford explained that the application of his mass/energy conversion would provide a better mechanism for heating the Sun than bird eyeballs and chariot wheels.

He stated:

> "The discovery of the radio-active elements, which in their disintegration liberate enormous amounts of energy, thus increases the possible limit of the duration of life on this planet, and allows the time claimed by the geologist and biologist for the process of evolution."

> Ernest Rutherford, professor of physics McGill University, Montreal

As evidenced by Rutherford's statement, the concept of evolution was an active participant in shaping the world of astronomy. Ernest Rutherford's model of a fission-powered Sun employed evolution theory as an evidence supporting his idea of a Sun fueled by fission – a Sun that could exist long enough to support Darwin's schedule of evolution.

The world of science now hinged on the assumptions of Copernicun gravity, flying photons, Cavendish Earth, gravity equality, 6.67, strange gases, random evolution, Darwin's timeline, neutral gravity, Helmholtz heat, constant light speed, Lorentz compression, flying electrons, and that mass

may be converted into energy contrary to finding of Michael Faraday.

The era of scientific theory during which the Sun was powered by fission lasted approximately 18 months. By then, spectroscopic analyses of the Sun were coming back.

Rutherford expected spectroscopic analysis to find radium, polonium, and heavy metals. (But not uranium because he didn't know about that.) Spectroscopic analysis did not agree with Rutherford at all. Spectroscopic results revealed hydrogen and virtually nothing else.

But so what?

If the Sun were composed of radium, polonium, or anything heavier than water, then would it not develop or hold such a massive atmosphere that you could not see what is below that atmosphere?

Hydrogen is the lightest of all elements. Isn't it obvious that hydrogen would remain on top, where it would be seen?

But that would be science by inference.

Science uses inference a lot, but, in this case, astronomers stuck with direct observation. The Sun, it was declared, is made of hydrogen and a little helium all the way through.

That method, logically, is like sending a probe to the moon to photograph the Earth in order to determine what Earth is made of. That flawed strategy would report that the Earth is composed of 70% water, 20% Alumina, 12% silica, and 6% iron because that's what's on the surface.

Just two years after Rutherford proposed metal stars, that theory was totally eclipsed by the concept of hydrogen-based stars.

The world of science now hinged on the assumptions of Copernicun gravity, flying photons, Cavendish Earth, gravity equality, 6.67, strange gases, random evolution, Darwin's

timeline, neutral gravity, Helmholtz heat, constant light speed, Lorentz compression, flying electrons, Faraday's folly, and that stars are hydrogen are made of hydrogen.

Relativity

In 1905 Henri Poincaré (1854 – 1912), a French mathematician, theoretical physicist, engineer, and philosopher of science, published a short paper in which he formulated the concepts of special relativity and space-time. Poincaré also translated Lorentz's contraction into special relativity terminology. Poincaré postulated the theory of the graviton and a theory of gravity waves.

Lorentz contraction

$$x' = \frac{x - vt}{\sqrt{1 - \frac{v^2}{c^2}}}$$

$$y' = y$$
$$z' = z$$

$$t' = \frac{t - \frac{v}{c^2} \cdot x}{\sqrt{1 - \frac{v^2}{c^2}}} \cdot$$

The key difference between Poincaré's relativity and Einstein's relativity is that Poincaré's relativity is an aether-space theory, whereas Einstein's is an empty-space theory.

Poincaré's aether-space-time was faulted as not predicting a universally constant speed of light.

Special Relativity

In 1905 Albert Einstein (1879-1955) used many of Hendrik Lorentz's equations, concepts, and techniques in his paper called "On the Electrodynamics of Moving Bodies". That paper become known as *Lorentz-Einstein Theory* and is now generally referred to as *Special Relativity.*

Einstein's specialty was exploring the theoretical physics of light.

Michelson-Morley's embarrassing failure of not getting a constant universal speed of light was seen by Albert Einstein as proving that no material aether existed. As for the speed of light, Einstein insisted that the speed of light was universally constant. Just like Michelson-Morley. His reasoning? Einstein decided that since Michelson-Morley's version of the aether was wrong, that there just wasn't any aether at all, thus returning to the philosophies of Isaac Newton.

The world of science now hinged on the assumptions of Copernicun gravity, flying photons, Cavendish Earth, gravity equality, 6.67, strange gases, random evolution, Darwin's timeline, neutral gravity, Helmholtz heat, constant light speed, Lorentz compression, flying electrons, and that space is composed of nothing.

What does that have to do with the speed of light?

Nothing.

1906
Nobel Prize in Physics

Joseph John Thomson
"in recognition of the great merits of his theoretical and experimental investigations on the conduction of electricity by gases."

But, since Einstein's light speed was identical to Michelson-Morley's Luminiferous Aether light speed – both universally constant – Lorentz contraction explained away certain paradoxes of Einstein's Special Relativity the same way it explained away the paradoxes of Michelson-Morley's LE.

Now Fitzgerald-Lorentz compression meant 'compression proportional to movement relative to the standard of space-time'.

1907
Nobel Prize in Physics

Albert Abraham Michelson
"for his optical precision instruments and the spectroscopic and metrological investigations carried out with their aid."

Thus, the most famous theory of Hendrik Lorentz was transformed. What had originated as an attempt to save aether theory was converted into a justification for the Einsteinian universe. Instead of becoming the champion of aether physics, Lorentz became a symbol of its downfall.

Max Planck protested against Einstein's theory of *flying photons* hearkening back to Isaac Newton for a dogged conclusion:

> "*The theory of light would be thrown back not by decades, but by centuries, into the age when Christiaan Huygens dared to fight against the mighty emission theory of Isaac Newton.*"

Planck saw Einstein's assertions as like Newton's *corpuscular theory of light*, and, worse, saw adopting flying photon theory as disregarding every discovery from Huygens through Maxwell – those being based on wave-physics.

Later attempts to mend particle- and wave-physics employed defining photon transmissions as 'waves of photons' – winds of waves of uncolliding photons blowing at 300 million meters per second in every direction simultaneously.

Wow.

A key sticking point concerning flying photon theory is the double-slit experiment. The double-slit experiments demonstrates that neither electrons or photons exist in transmission. If photons do not exist in transmission, then how is light transmitted?

A key sticking point concerning wave theory is equally disturbing. If light exists as waves, is transmitted as waves, and behaves as waves, then what are photons?

WAVE-PARTICLE TRANSFORMATION (WPT)

To define light as either a wave or a photon is paradoxical. Light – in various situations and interactions – assumes the attributes of a wave some points, and the attributes of photons at other points.

Einsteinian physics essentially shrug at this problem. Instead of observing this as impossible, Einsteinianists ask, "How does it do that?" Light, as flying photons, is assumed to bridge this paradox in a way not yet observed, and the paradox is softly referred to as 'wave-particle duality'.

The paradox to address is that light-waves and photon particles obey different physical laws. How can light be both? Waves have no mass. Photons are particles which, within conventional physics, must have mass. How can light be both? The answer is: It never is.

For photons to behave as photons, they must be photons. For light waves to behave as light waves, they must be light waves. And for light waves to cause photons, and for photons to cause light waves, there must be transformation.

Wave-particle transformation occurs when light waves affect an atomic structure. When light waves affect the aether-pi* field near an atomic structure, photons and/or electrons are induced and/or destroyed.

*aether-pi is new concept of aether. More on aether-pi later in this book.

The energy of light is never a wave and a particle. It is sometimes a wave and sometimes a particle. Particles occur *from the medium*. A similar phenomenon occurs in the vicinity of high-energy detonations. Near an explosion, as the high-pressures are transmitted to the atmosphere, the atmosphere compresses so much that some gases condense into liquid particles. As the shock-wave passes, the condensed particles immediately evaporate.

The same thing happens in the high-pressure regions surrounding supersonic airplanes, forming a cloud around them.

United States Navy
photo ID 061105-N-8591H-389

This cloud is not flying with the airplane, but this cloud is always around the airplane. This cloud forms after the airplane arrives and dissipates before the airplane leaves.

Similarly, when a light-wave affects an atom, a cloud of photons forms around the atom. As the light waves passes, the photons 'evaporate' or explode. Where did the energy go? The light-wave – the energy and momentum of the light wave being the airplane in the former example – continues on *mostly.*

Some of the energy is deflected, so entropy occurs and, in the airplane example, that loss of energy slows the airplane down. The thrusters must keep pushing the airplane forward to make up for that loss of energy.

THE LAW OF WAVE-PARTICLE TRANSFORMATION

input = output

This is far more important than its obviousness implies.

For a wave to induce/transform into a photon, and also for a photon to cause/transform into a wave, and to do so with no losses, indicates perfect transformation.

Typical electromagnetic transformers lose 20% of the energy transformed to entropy. You put $1.00 of electricity in and you get $0.80 of electricity out.

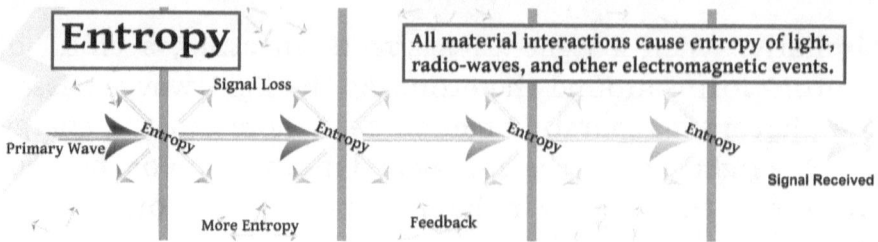

For waves to be losslessly transformed into photons, the field where those waves occur must be incapable of capacitating those charges. If that were not true, than light would not transform at 100% efficiency. Instead, it would entropy, 'wear out', fade, and diminish without material interaction.

If light experienced entropy in consequence of transmission, the Hubble telescope would not see or detect galaxies 13 billion parsecs away.

But light waves do not entropy.

Light waves are entropy.

The distance a light-wave may be transmitted absent of material interaction is limited only by the intensity of that wave. When the charge of that wave is insufficient to bias the medium transmitting that wave at a Planck scale, that light wave becomes extinct.

So no. Light does not go on forever.

For instance, the light from an intense source like a star obviously makes it 13 billion parsecs to Earth, but the weak light of a Pluto-sized planet that far away might have become extinct at a shorter distance, so we might be physically unable to see a 'Pluto' within one of those galaxies.

WAVE-PARTICLE TRANSFORMATION AND PHOTOSYNTHESIS

Photosynthesis occurs when light waves affect a photosynthetic chemical in the presence of water and carbon dioxide. When that happens the photons induced by the light waves are focused to rearrange two stable chemicals – H_2O and CO_2 – into unstable chemicals – volatile organic compounds (VOCs), such as $C_6H_{12}O_6$ and $C_{12}H_{22}O_{11}$, and oxygen compounds, such as atmospheric oxygen which is O_2 and ozone which is O_3.

The process of dividing and reforming chemicals involves heterolytic fission (heterolysis) and/or homolytic fission (homolysis). Heterolytic fission occurs when a chemical or allotrope divides into two atoms, leaving one atom with a positive (+) charge and the other atom with a negative (-) charge. Homolytic fission occurs when a chemical or allotrope is divided into atoms having the same charge.

Heterolytic and homolytic fission are a close-up view of what causes chemical reactions. The imbalances present after either one leads to other chemical reactions.

The energy necessary to generate *chemical fission* is called the bond-dissociation energy (BDE, D_0, or $DH°$) and those values are quite high for stable chemicals. Water has a BDE of 926 kJmol. Water chemically fissions (disassociates) into hydrogen and oxygen at $2,273°$ Kelvin or $3,632°$ Fahrenheit.

That is just a taste of how much energy it takes to make water into sugar. Next you have to add some CO_2. CO_2 has a BDE of 1,598 kJmol.

A photosynthetic chemical develops enough power to rearrange these chemicals by the dozen. Literally. Twelve CO_2s and twelve H_2Os or six of each for a total of twelve.

Effectively, each C attaches to one H_2O and releases an O_2. That results in a positive-bias chemical – CH_2O – and a negative-bias chemical – O_2 – in sets of 6 or 12.

The energy of the light-waves affecting the photosynthetic reaction have been centralized as energetic particles – photons – and those photons' electromagnetic potential is used to break apart stable chemical structures – water and CO2. Those electromagnetically charged particles are then used to 'weld' new chemical structures together.

The photon charges then remain as permanent parts of new chemical structures either as ambient potential or expressing as electron charges or chemical bonds.

Photons can remain locked up this way – within chemical structures – for billions of years. A very important implication of this is that photons do not travel. Photons are localized physical particles which exist near or within atomic and chemical structures.

Positive (+) photon charges have been added to the CH_2O (or complex thereof) and negative (-) photon charges have been added to the O_2. Due to this bipolarity, these chemicals physically, magnetically attract to each other.

Due to this attraction, they tend to collide.

At any future point in time, if and when the sugar or VOC comes in contact with atmospheric oxygen, the two chemicals will react. The sugar burns or, at low temperatures, catalyzes with the oxygen and that breaks the chemicals apart. The photon energy is released as light-waves as the elements then rearrange into stable chemical structures – water and CO_2.

> A 'stable' chemical is any chemical which is neutral relative to the chemicals in its vicinity. Hydrochloric acid, for instance, is stable among hydrochloric acid, but it is unstable in the vicinity of metals.

Photons are induced by light waves. Photons contain (capacitate) the energetic potential of light waves. When photons explode or become destroyed, the energetic potential formerly retained as a photon is released as light waves.

If enough photon energy is released fast enough, material photon populations skyrocket, resulting in plasma (fire) and very high temperatures. Those very high photon populations (temperatures) cause chemical fission and even atomic fission. This may result in either explosion, a persistent fire, or an atomic fission reaction depending on the resultant chain reaction.

Since sugars and other carbohydrates contain so much oxygen, carbohydrates can burn without the presence of atmospheric oxygen. This is done to generate charcoal.

To create charcoal:

- dig a hole larger than 2 feet in diameter and at least 3 feet deep;

- nearly fill the hole with wood;
- ignite wood;
- allow to burn until embers form (the glowing middle that keeps the fire going after it's established);
- cover hole (exhaust will occur);
- wait eight hours or until exhaust ceases;
- allow to cool; remove charcoal from hole.

You have burned the oxygen out of carbohydrates.

Unlike wood, charcoal is not a carbohydrate. Charcoal consists of hydrocarbons left behind after the oxygen burned away. The oxygen burned away with hydrogen.

$$C_6H_{12}O_6 - 6*(H_2O) = C_6$$

If the chemicals were pure and the reactions were perfect, you would have pure carbon left over – called *graphite.*

The dry charcoal left behind is flammable in the presence of atmospheric oxygen. It still contains as much as 50% of the photon/heat potential which was present in the wood.

Wet charcoal is flammable *even without* the presence of atmospheric oxygen, but difficult to ignite. At around 2,273° Kelvin or 3,632º Fahrenheit, water chemically fissions, providing oxygen to burn the charcoal and releasing hydrogen towards outer space.

$$C_6 + 12*(H_2O) = 6*(CO_2) + 12*(H_2)$$

In 1908 Wilhelm Wien discovered a method deflecting electromagnetic particles using magnetic fields. This is the technology used in *polarized lenses*. Polarized lenses used *doped* glass which is magnetically polarized when cooling. This polarization alters how light is conducted through the glass.

This phenomenon demonstrates that wave-particle trans-formation occurs relative to magnetic fields.

In 1908 Johannes Wilhelm "Hans" Geiger (1882 – 1945) developed the *Geiger Counter* – a device which measures high-frequency radiation such as x-rays and gamma-rays.

Nuclear Gold

In 1908 the *Geiger–Marsden experiments* began. While the technique is Rutherford's, the most famous experiment using his technique was not performed by Rutherford, but was carried out under Rutherford's direction.

1908

Nobel Prize
in Chemistry

Ernest Rutherford

"for his investigations into the disintegration of the elements, and the chemistry of radioactive substances."

The most famous gold-foil experiment series was performed by Hans Geiger and an associate and was called the *Geiger–Marsden experiments.* The Geiger-Marsden experiments began in 1908 and ran into 1913.

Through this technique, it was demonstrated that gamma radiation was not deflected by gold, but alpha particles (helium nuclei) *were* deflected by the gold foil. Rutherford showed that this demonstrates that atoms have a central point – a nucleus.

Rutherford inferred that that is where the mass of the atom must be – in the nucleus. With the mass in the nucleus, Rutherford proposed that electrons were like little planets orbiting atomic nuclei.

The world of science now hinged on the assumptions of Copernicun gravity, flying photons, Cavendish Earth, gravity equality, 6.67, strange gases, random evolution, Darwin's timeline, neutral gravity, Helmholtz heat, constant light speed, Lorentz compression, flying electrons, nothing space, orbital electrons, and the idea that mass is concentrated inside of an atomic nucleus Copernicun style.

In 1908 at Leiden University in the Netherlands Willem de Sitter (1872 – 1934) was appointed to the chair of astronomy. De Sitter originated the Einsteiniverse concepts of *de Sitter space* and *de Sitter universe* in 1932. I'll explain that when we get there.

Slipher Expansion

1908
Nobel Prize in Physics

Gabriel Lippmann
"for his method of reproducing colours photographically based on the phenomenon of interference."

In 1912, Vesto Melvin Slipher (November 11, 1875 – November 8, 1969), an American astronomer, spectrographically determined spectral lines of galaxies. He deduced and calculated that many galaxies had significant red-shift values.

1911
Nobel Prize in Chemistry

Marie Curie, née Sklodowska
"in recognition of her services to the advancement of chemistry by the discovery of the elements radium and polonium, by the isolation of radium and the study of the nature and compounds of this remarkable element."

Slipher completed his doctorate at Indiana University in 1909 and moved to Flagstaff, Arizona where he spent the rest of his career at the

Lowell Observatory, and the rest of his life gazing up into those clear Arizona skies.

1911
Nobel Prize in Physics

Wilhelm Wien
"for his discoveries regarding the laws governing the radiation of heat."

By 1912, Slipher had established spectral signatures for distant galaxies.

Origin of Continents

In 1912 Alfred Lothar Wegener (1880 – 1930) – a geophysicist and meteorologist born in Berlin – proposed that Earth's continents were drifting about. He detailed these theories in 1915 in a book titled *The Origin of Continents and Oceans*.

Wegener originally proposed that this drift was due to tidal action and centrifugal force. He later changed his position.

In 1929 Wegener posited that continental drift resulted from subterranean convective actions.

He may have been right the first time.

Either way, this became the beginning of plate tectonic theory. Wegener proposed that Earth's continents were formerly connected to Antarctica. This was evidenced in large part, he claimed, through the distributions of various types of fossils.

Wegener coined a phrase for the land-form from which all continents spread – Pangaea. Pangaea is a Greek word which means 'Entire Earth'. According to how one interprets the term *earth*, that can mean one of of two things.

If he meant 'entire Earth' – as 'entire planet', then he meant that Pangaea covered Earth's whole surface.

If he meant 'entire earth' – as 'all of the dirt', then he meant that Pangaea included all land masses, but not the entire planet.

Wegener based his Pangaea/evolution model on the fossils of land-dwelling creatures.

Why not fish?

Because fish are everywhere. There are fish and crustacean fossils near the peak of Mount Everest. Not just in Mount Everest, but in all mountain ranges.

Many of the peaks and plateaus of the Rocky Mountains were carved with water and tidal action including Colorado's Garden of the Gods and Utah's Arches National Park.

In South America there are massive salt flats exceeding 14,000 feet in altitude. Salt flats occur when an oceanic (saltwater) body becomes cut off from the ocean due to falling ocean levels. When the water dries out, the salts become concentrated.

America's inland ocean used to cover Bonneville salt flats and included Utah's Great Salt Lake. The Great Salt Lake is now

4,200 feet above sea level.

For water to permanently exist at those levels for millions of years would require exponentially more water than exists on, in, or near planet Earth. Water levels that high are impossible if Earth has always existed at its present radius or if it has always had mountains.

Wegener was not the first man to assume that Earth's radius has remained constant, but with Wegener that concept became codified.

The world of science now hinged on the assumptions of Copernicun gravity, flying photons, Cavendish Earth, gravity equality, 6.67, strange gases, random evolution, Darwin's timeline, neutral gravity, Helmholtz heat, constant light speed, Lorentz compression, flying electrons, nothing space, orbital electrons, Copernicun mass, floating continents, and constant planetary size.

In 1913, Crookes invented UV-blocking sun-glass lenses by *doping* glass with *cerium* – a silvery-white lanthanide metal. The cerium-doped glass reflected 100% of UV light and 90% of infrared light with only slight visual tinting.

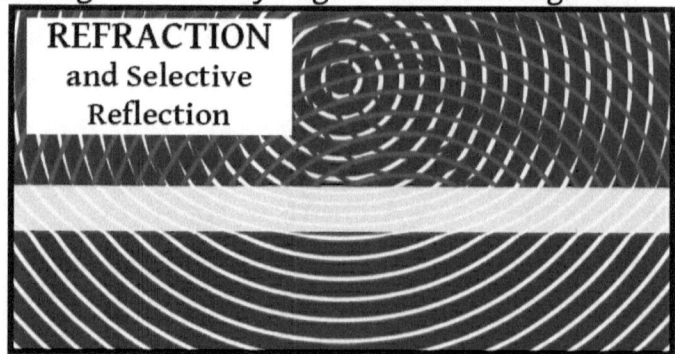

REFRACTION
and Selective
Reflection

Surface treatments such as window tinting also have this effect. A treated window may reflect incoming ultra-violet radiation while letting yellow light through as shown in above.

Treatments or lenses with treatments are sometimes called

filters. Filters can be used to block red light, blue light, or other frequencies. Filters may be very, very dark allowing telescopes to be aimed directly at the Sun without damaging sensitive equipment.

In 1914, Slipher established that spiral galaxies rotate.

1914

Nobel Prize in Physics

Max von Laue

"for his discovery of the diffraction of X-rays by crystals."

In 1914 Max Planck became dean at the University of Friedrich Wilhelm (Berlin) University and promptly exercised his new powers to create a professorship position for Albert Einstein.

Arthur Eddington

The man who became Albert Einstein's greatest supporter was Arthur Eddington (1882-1944).

Eddington is famously known for the *Eddington Limit* – a theoretically natural limit to the luminosity of stars based on stellar fusion theory.

Concepts which violate the Eddington Limit (more gently referred to as *Eddington Luminosity*) include gamma-ray bursts, novae, supernovae, x-ray binaries, active galaxies, neutron stars, white dwarfs, and black holes.

The world of science now hinged on the assumptions of Copernicun gravity, flying photons, Cavendish Earth, gravity equality, 6.67, strange gases, random evolution, Darwin's timeline, neutral gravity, Helmholtz heat, constant light speed, Lorentz compression, flying electrons, nothing space, orbital electrons, Copernicun mass, floating continents, constant planet size, and Eddington limitocity.

General Relativity

In 1915 Einstein published four groundbreaking new papers which established what would later become known as: *The General Theory of Relativity.*

Einstein identified light as a locally constant phenomenon proportional to mass and energy. Although Einstein thought light consisted of flying photon, Einstein's theories and famous equations have been heralded as central to phenomenal advancements in the theoretical and physical sciences.

Special Relativity was special in that it referred only to light. Later, *General Relativity* was designed to address physics generally. General relativity included the concepts of gravity and time.

Nothing is Nothing Paradox

In designing general relativity, Einstein ran into some logical problems. His space was composed of nothing. Nothing cannot have any attribute. Nothing is just nothing. But Einstein realized that his nothing had effects on time and gravity, even bending light. So Einstein assigned his nothing a whole host of attributes and called it space-time.

This renders Einstein's whole theoretical structure paradoxical. Now infinite amounts of nothing have measurable effects on everything and the only rule in the universe is the speed of light.

Adding Lorentz transformation to Einstein's space-time was like throwing two paradoxes in a blender. Now nothing – which was made of nothing – also made objects flatten whenever they moved relative to nothing.

So Einstein designed an aether theory in which the aether is referred to as 'nothing' and the aether is named 'space-time'.

The world of science now hinged on the assumptions of Copernicun gravity, flying photons, Cavendish Earth, gravity equality, 6.67, strange gases, random evolution, Darwin's timeline, neutral gravity, Helmholtz heat, constant light speed, Lorentz compression, flying electrons, nothing space, orbital electrons, Copernicun mass, floating continents, constant planet size, Eddington limitocity, and that the nothing of space has infinite powers of gravity, time, and energy.

On May 25, 1919, Arthur Eddington led an expedition to Príncipe Island in West Africa, where he saw the evidence he sought by observing an eclipse.

What he was looking for was the 'flattening of stars' near the horizon of the Sun. Einstein had predicted that photons have mass. Based on that, Einstein predicted that photons would change direction slightly when passing near the Sun. Einstein said that would cause stars seen over the horizon of the Sun and appear flattened.

Eddington claimed that he witnessed the flattening of stars. But not directly. In fact, it had been a cloudy day on Príncipe Island. Eddington based his conclusions on photographs from a partner expedition in Brazil.

Under future scrutiny, Eddington's 1919 findings were "too inaccurate" to demonstrate his theory. Nevertheless, Eddington's 1919 finding are still popularly reported – 100 years later – as one of three 'fundamental proofs' of Einstein's relativities – 'photons have mass'.

Eddington's conclusions created yet another paradox. If light travels as flying photons, and photons have mass, then according to e=mc², photons have infinite mass.

The world of science now hinged on the assumptions of Copernicun gravity, flying photons, Cavendish Earth, gravity equality, 6.67, strange gases, random evolution, Darwin's timeline, neutral gravity, Helmholtz heat, constant light speed, Lorentz compression, flying electrons, nothing space, orbital electrons, Copernicun mass, floating continents, constant planet size, Eddington limitocity, omnipotent nothing, and photons with mass flying around at the speed of light

destroying the whole universe because $e=mc^2$ says so.

The Age of Paradoxes was now in full swing. Scientists would spend the next 100 years as mathematical philosophers pondering the meanings of infinities.

Infinities meant black holes, worm holes, singularities, big bangs, anti-particles, anti-matter, dark matter, dark energy – a whole carnival of impossible paradoxes woven together with the fabric of space-time which would be paraded before the world as fact.

It was *Barnum and Bailey's Sci-fi Circus of Physics* complete with magical particles, disappearing tricks, invisible forces, a universe popping out of nothing, high-wire mathematical tricks, and matter bending like a clown in a funny mirror. A circus of ideas that no natural man can comprehend.

But it was not a hoax.

One hundred years ago, those were the best theories we had.

Solar Fusion

In 1920, using Einstein's (or Rutherford's) mass/energy proportionality and Aston's hydrogen/helium inequality (which Aston had not yet published) Eddington, from Trinity College, Cambridge along with Rutherford, Aston, and Blackett proposed in the paper *"The Internal Combustion of Stars"* that fusion of hydrogen into helium is what powers solar and stellar discharges.

Francis Aston's findings were reported two years later.

Using the concept of mass/energy proportionality, which Einstein expressed as $e=mc^2$, Eddington calculated that if the Sun was made of hydrogen, and that that hydrogen fuses to helium, that the resulting helium has less mass and, therefore, energy has been expressed.

Eddington proposed that hydrogen clouds could condense on themselves, that the Kelvin–Helmholtz mechanism could get fusion reactions started, and then that those fusion reactions could become self-sustaining.

The Kelvin-Helmholtz mechanism, however, does not generate high internal temperatures. It generates <u>high external</u> temperatures *because* the inside gets cold. Heat and density are counterparts. More of one means less of the other.

Fusion is not as easy as it sounded back then.

The world of science now hinged on the assumptions of Copernicun gravity, flying photons, Cavendish Earth, gravity equality, 6.67, strange gases, random evolution, Darwin's timeline, neutral gravity, Helmholtz heat, constant light speed, Lorentz compression, flying electrons, nothing space, orbital electrons, Copernicun mass, floating continents, constant planet size, Eddington limitocity, omnipotent nothing, infinite photons, and stellar fusion.

In 1920, Eddington published *Space, Time and Gravitation.*

Before Eddington's stellar fusion got off the press, trouble was brewing. Scientists were developing technologies to view the Sun. The uses of these technologies included the ability to look into a solar hole. Earlier attempts to view the Sun often damaged sensitive equipment.

Solar holes are intermittent dark, low-temperature regions of the Sun which typically occur in the middle latitudes. Solar holes, which were first noted by Galileo, occur when cloud layers are missing or dissipate, revealing lower-altitude regions of the Sun.

New technologies allowed scientists to view the 'surface' of the Sun, and it was found to be less than 10,000° Kelvin. That's not hot enough for fusion. That's not nearly hot enough.

Modern observations observe the surface of the Sun as 5778° degrees Kelvin. The minimum temperature for fusion of hydrogen (theoretical) is 13° million Kelvin. There's no fusion on the Sun.

This paradox of fusion theory intermittently re-enlivens stellar fission theory. Stellar fission *could* occur at those temperatures. Further, fission could form hydrogen and helium from heavier elements, thus explaining not only the presence of those lighter elements, it could also explain the solar wind.

If heavy element disintegration was constantly producing hydrogen and helium, then those gases would be exhausted from the Sun, forming a 'solar wind'. The fusion model of the Sun predicts the opposite – it predicts that hydrogen and helium must flow <u>into</u> to the Sun to supply or 'fuel' fusion.

In 1921, Enrico Fermi (1901 – 1954), born in Italy, published *On the dynamics of a rigid system of electrical charges in translational motion*. This treatment of Einstein's relativities treated mass in a strange, new way. Instead of treating mass as being a constant value, it treated mass as a *tensor* – a variable.

1921
Nobel Prize in Physics

Albert Einstein

"for his services to Theoretical Physics, and especially for his discovery of the law of the photoelectric effect."

This spawned the concept of *relativistic mass* whereby mass *gains* mass in consequence of movement relative to some universal standard. In this model, when an object accelerates, the mass of that object literally increases. The faster the aetherflow, the greater the mass.

Einstein's relativistic mass is calculated by dividing the dividing the mass of the object in motion by the square root of the opposite of the velocity squared divided by the speed of light squared as shown below.

$$\text{Relativistic Mass} = \text{Mass} / (1 - \text{velocity}^2 / c^2)^{(1/2)}$$

Below, the speed of light is shown in blue and the relativistic mass is shown in red.

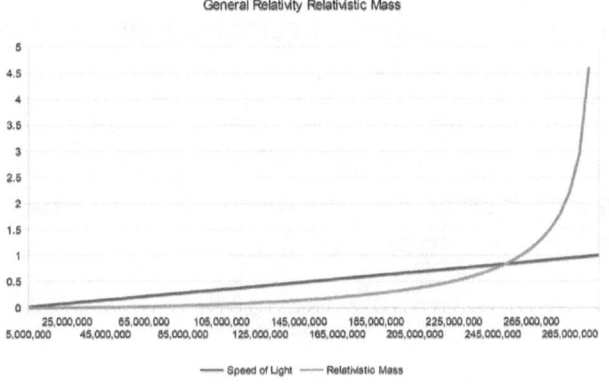

General Relativity Relativistic Mass

According to this equation, when an object is going half of the speed of light, its mass increases by about 16%, at 87% of the speed of light the mass has doubled, and once an object reaches the speed of light its mathematical mass becomes infinite.

In 1921 Niels Henrik David Bohr (1885 – 1962) opened the Niels Bohr Institute. Bohr, a molecular the theoretical physicist, worked with JJ Thomson, Ernest Rutherford, William Bragg, and others.

Bohr developed the *Bohr model* of the atom, where a nucleus holds orbiting electrons in place the way the Sun holds

planets in orbit. The Bohr model is now considered foundational but obsolete.

Aston's Isotopes

In 1922 Francis William Aston (1877 –
1945), an chemist and physicist born in
Harborne, Birmingham, England, showed
the world that, as masses go, four hydrogens don't add up to one helium.

1922
Nobel Prize
in Chemistry

Francis William Aston
"for his discovery, by means of his mass
spectrograph, of isotopes, in a large
number of non-radioactive elements,
and for his enunciation of the
whole-number rule"

After returning from serving his country
in World War I, Aston built his first *mass
spectrograph* in 1919. Aston's improved mass spectrograph
divided isotopes from chemicals. Before 1921, Aston had
identified over 212 naturally occurring isotopes, ushering in a
new era of atomic science.

Among those 212 isotopes existed the data Aston is best
known for – the data indicating that 4 hydrogen atoms weigh
more than 1 helium atom - the data Eddington had employed
to explain solar emissions - the basis of *stellar fusion theory.*

Fusion Chain Paradoxes

Advances in extreme-condition experimental testing have
worsened the paradoxes of stellar fusion theory. Hydrogen
now requires 13 million Kelvin to fuse. When hydrogen fuses,
it (theoretically) emits radiation. During fusion, conditions
produced may (theoretically) climb to 26 ° million Kelvin.

But hold on. This theory forgot something. When hydrogen
goes ultra-critical and becomes atomically unstable, it does
not go jump in bed with other hydrogens. It explodes instead.
If fissions.

And when light elements – elements lighter than iron – fission,
there is a negative energy output. The reaction is *endother-*

mic. The ultra-critical destruction of a hydrogen atoms cools things off.

With all that fission, temperatures will never get high enough to disturb helium atoms. That requires temperatures of 100° million Kelvin. And if temperatures rise to 100° million Kelvin, helium fissions away. Basically, in order for anything to experience fusion, it must already be in the process of fission.

Next on the list is carbon. If maybe somehow some of that helium became carbon, that carbon is stable to 500° million Kelvin. This is where the real problems begin.

Carbon won't fuse to carbon, so you have to add carbon to helium. To do this, you need carbon ready to fuse at 500° million Kelvin and helium (which fissioned at 100° million Kelvin) right there with it.

The chances of carbon at 500° million Kelvin being the same place as helium at 100° million Kelvin are null.

That being thoroughly impossible, things only get worse from there. Soon you get to iron. Anything bigger than an iron nucleus is *endothermic* when fused. So if iron begins experiencing fusion (with all the same impossibilities formerly listed) then temperatures will fall, and soon it won't be hot enough to fuse anything. Before things get hot enough for fusion, fission cools things off.

Stellar fusion theories are logically impossible, mathematically impossible, and constitute one big ball of flaming paradoxes.

Arthur Holly Compton (1892-1962) was an American physicist born in Ohio best known for his research into electromagnetic effects on materials and the *Compton Effect*.

The Compton Effect observes that when light goes into an object, that different light comes out. This is most obvious when high-energy light is used, such as gamma rays or x-rays.

Compton observed that when light is reflected from a flat surface it does not all reflect in the same direction.

Instead, from the point where the light arrives, *most* of the light is reflected but *some* of the light leaves at completely different angles.

Would a light wave do that?

Well, yes, actually. As seen here, if an atomic structure is not flat and light waves are reflected from that surface, then those waves will not all reflect in the same direction.

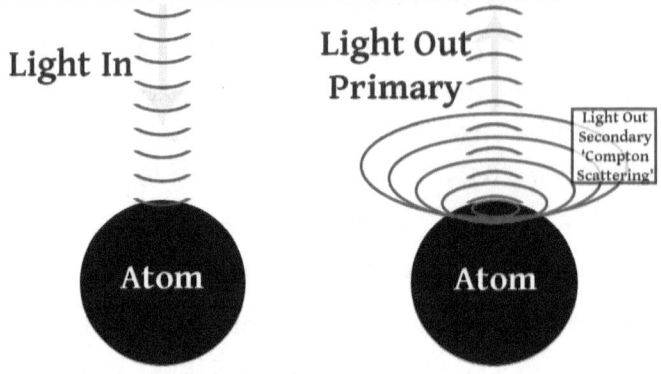

Russian Expansion

Alexander Alexandrovich Friedmann (1888 – 1925) was a Russian Soviet physicist and mathematician famous for the *Friedmann equations*. Friedmann was the son of a ballet dancer and a piano player. His mother was the piano player.

In 1922, Friedmann developed the theory of an expanding universe and shared his findings with a Belgian astronomer named Georges Lemaître at a conference in Copenhagen.

In 1923, Compton furthered his concept of the Compton Effect by demonstrating that when light-waves were affecting an atomic structure that apparently *inertial* interactions occurred.

Compton had also determined that the angle of the path of

light could also be manipulated using electromagnetic fields. Heralding back to the cathode ray experiments – which were interpreted to suggest that electrons had mass – Compton likewise suggested that the bending of light within an electromagnetic field demonstrated that photons had mass.

Science is the Art of Interpretation

Eddington: More Einstein than Einstein

Also at Cambridge in 1923, Arthur Eddington released what some consider his greatest work: *The Mathematical Theory of Relativity*. Albert Einstein himself stated that that was the finest presentation of the subject of relativity in any language. The work included original mathematical contributions by Eddington, making it distinctly his.

The Mathematical Theory of Relativity made Eddington a leader in the field of relativity physics.

In 1923, Georges Henri Joseph Édouard Lemaître (1894 – 1966), a Belgian Jesuit Roman Catholic Priest, mathematician, and astronomer, became a research associate in astronomy at Saint Edmund's College at Cambridge University.

Lemaître, a World War I veteran and ordained priest with a doctorate in physics, crossed paths with Arthur Eddington there. It was there that Lemaître set his own sights on general relativity.

In 1924, in the German physics journal *Zeitschrift für Physik*, Friedmann published *"On the possibility of a world with constant negative curvature of space"*, or, in other words, 'If the universe is expanding...' - a proposal which is popularly accepted as fact in the modern world and a second precursor to the theories of Lemaître and Hubble.

Hubble's Universe

Meanwhile, on November 23, 1924, Edwin Powell Hubble (1889 – 1953) published in the *New York Times* that spiral nebulae including the Andromeda Nebula and Triangulum were not nebulae, but vastly larger objects – distant galaxies far, far away in a time long, long ago.

Building on the work of Slipher and others, Hubble's findings were formally reported to the astronomical community in 1925, and published in peer-reviewed form in 1929. That is when Slipher's concepts began to become generally accepted.

The world of science now hinged on the assumptions of Copernicun gravity, flying photons, Cavendish Earth, gravity equality, 6.67, strange gases, random evolution, Darwin's timeline, neutral gravity, Helmholtz heat, constant light speed, Lorentz compression, flying electrons, nothing space, orbital electrons, Copernicun mass, floating continents, constant planet size, Eddington limitocity, omnipotent nothing, infinite photons, stellar fusion, and the Russian theory of an expanding universe.

The Philosopher's Stone

In 1925 Patrick Maynard Stuart Blackett (1897 – 1974) became the first person to prove that radioactivity could cause nuclear transmutation of one element into another.

Blackett worked with Aston, Rutherford, Eddington, Geiger, and others famous physicists at Cambridge University, becoming a Fellow of King's College at Cambridge in 1923.

In 1927 Nikolai Konstantinovich Koltsov (1872 – 1940) proposed that hereditary traits were passed via a "giant hereditary molecule" made up of "two mirror strands that would replicate in a semi-conservative fashion using each strand as a template".

Quantum: How Much?

In June of 1926, Paul Adrien Maurice Dirac (1902 – 1984), born in Bristol, England, finished his PhD at St. John's College, Cambridge University with a thesis – the first thesis of its kind – titled *Quantum Mechanics*. Quantum mechanics is the science of mathematically quantifying atomic and subatomic physics based on units of light or *photons* – originally called 'quanta' – the smallest unit of measure.

A French schoolteacher's son, Dirac was a poor man. Dirac earned a scholarship to Cambridge in 1921, but did not have enough money to live on campus. Dirac became one of the few people who studied there anyway, attending for two years before an additional scholarship covered his housing.

Dirac finished first in his class.

"This balancing on the dizzying path between genius and madness is awful."

- Albert Einstein
concerning Paul Dirac

Dirac notoriously spoke very deliberately and slowly as he carefully matched words to ideas before speaking. This jokingly resulted in a physics unit known as the *Dirac.* One Dirac is equal to one word per hour.

Dirac's accomplishments and breakthroughs include the discovery of *fermions* and *fermionic matter* while working with Enrico Fermi. Dirac and Fermi also predicted the existence of anti-matter.

The world of science now hinged on the assumptions of Copernicun gravity, flying photons, Cavendish Earth, gravity equality, 6.67, strange gases, random evolution, Darwin's timeline, neutral gravity, Helmholtz heat, constant light speed, Lorentz compression, flying electrons, nothing space, orbital electrons, Copernicun mass, floating continents, constant planet size, Eddington limitocity, omnipotent nothing, infinite photons, stellar fusion, Russian expansion, and that everything in the universe has a doppelgänger opposite.

Particles and Anti-Particles

Anti-particles are simply electromagnetically inverted versions of particles.

All particles are formed from *aether-pi* – fundamental sub-atomic particles. The *charge* of a particle relates to the orientation of the aether-pi forming that particle. Some particles present a negative bias (negative magnetic surface charge), others present a positive bias (positive magnetic surface charge).

Natural bosonic matter is all positively charged. Negatively-charged bosonic matter is antimatter.

When a proton's charge is reversed, it displays a negative charge instead of a positive charge and is then called an anti-proton.

Oppositely-charged particles are magnetically attracted to each other, so anti-protons and ordinary protons are magnet-ically attracted to each other and those particles tend to col-lide.

When matter collides with anti-matter, the oppositely-charged nuclei scatter each other's inertial frames of refer-ence, and the particles cause each other to fission.

This does not result in energetic output. Fission of light elem-ents is *endothermic* requiring more energy to be invested in causing the reaction than the energy emitted from the reac-tion.

Anti-matter does not naturally exist. Anti-matter cannot exist in vicinity of ordinary matter.

In 1927 Geiger worked with Arthur Compton using the Geiger counter to confirm the existence of the *Compton Effect*. Using a Geiger counter.

1927
Nobel Prize in Physics

Arthur Holly Compton
"for his discovery of the effect named after him"
Charles Thomson Rees Wilson
"for his method of making the paths of electrically charged particles visible by condensation of vapour."

A Geiger counter is a very high-frequency radio antenna which collects x-ray and/ or gamma ray energy into electrical charges and emits clicks or shows radiation levels on a guage.

Compton witnessed Wave Particle Transformation (WPT) and Atomic Energy Transformation (AET). Atomic Energy Transformation is covered in more detail later. Compton interpreted these effects through Rutherford's paradigm of electrons orbiting atomic nuclei. He also employed Einstein's concept of empty space-time and Newton's concept of corpuscle light. Using these frames of reference, Compton concluded that light existed as flying photons. He concluded that those photons were interacting with orbital electrons.

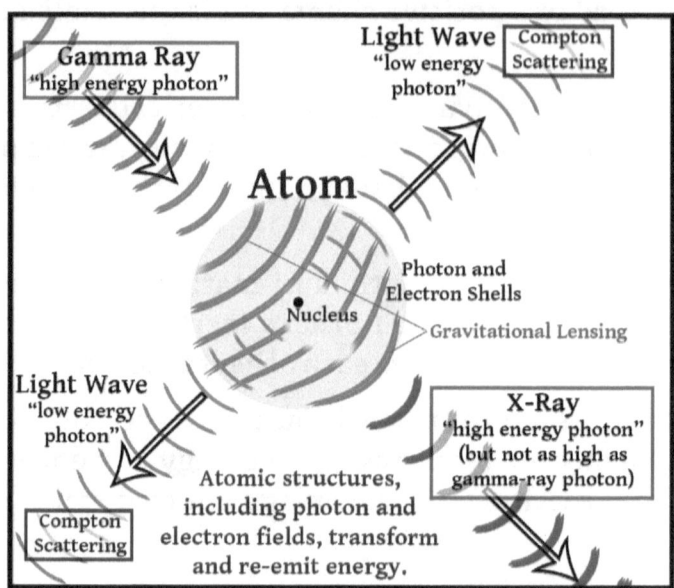

To account for the *uncertain* departure angle of various photons, photons were described as interacting with electrons, with photons being emitted when electrons fell to lower orbits.

This gave rise to a never-popular notion that photons orbit electrons.

The notion that photons are emitted in consequence of electrons falling to lower orbits raises some interesting questions.

- If electrons are persistent (bosonic) subatomic particles, why do they interact with photons?
- If electrons lose energy when photons get emitted, are electrons made of photons?
- If electrons orbit atoms, why don't photons orbit atoms?

- And since the double-slit experiments proves that photons and electrons do not exist in transmission, where do they exist?

There is also a whole other class of problems: Electrons are always negatively (-) charged. Photons are not.

- What happens when positive (+) photons interact with electrons?
- Do they destroy them? Reduce them?
- If the positive photons get destroyed, is there still light when a negative photon goes somewhere?
- Can light permanently strip electrons from atoms?
- If electrons interact with photons, then what does the nucleus do?

The Compton Effect gets a little stranger when the energy applied is too small to induce 'photon-sized' Compton scattering. Here, like in the double-slit experiment, light is demonstrated to be transmitted in the form of waves *even if those waves are <u>not</u> powerful enough to induce photons.*

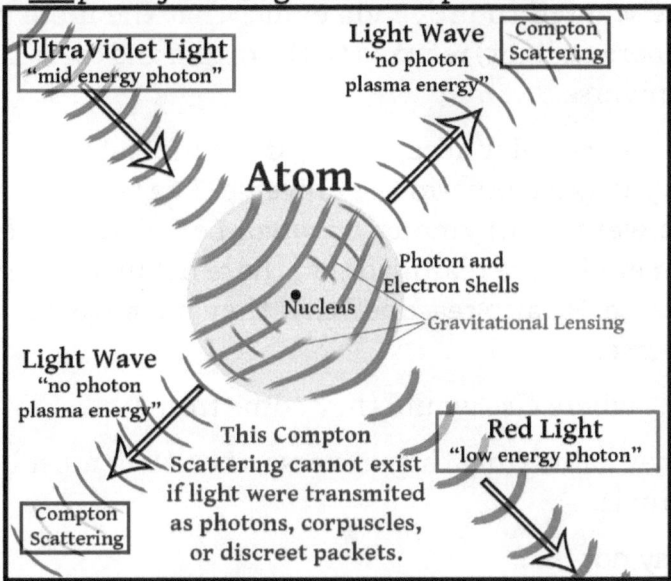

As with the cathode ray experiment, the Compton effect seems to indicate that light does not exist as flying photons and that light *in transmission* does not exist as photon-sized packets.

Origin of Earth

Also at Cambridge, Lemaître had realized that since space-time was composed of nothing, it had unusual flexibility. It could be compressed into nothing.

And with $e=mc^2$, matter consisted of energy. Energy is not three dimensional so that, too, can be compressed into nothing.

So Georges Lemaître found the nothings in Einstein's equations and beat them with a Bible.

In 1927 Lemaître published "A homogeneous Universe of constant mass and growing radius accounting for the radial velocity of extra-galactic nebulae" dues ex machina.

Lemaître wanted to prove via $e=mc^2$ that the Bible (Genesis 1:4 Let there be light) referred to the origin and creation of the whole universe.

Prior to the popular acceptance of Edwin Hubble's findings, the Milky Way galaxy *was* the universe. The terms galaxy and universe were synonymous. So when Lemaître found 'extra-galactic nebulae', he reasoned that those nebulae were beyond the universe. He also reasoned that they must have *come from* the universe.

Because... where else would they come from?

And if they came *from* this universe, then they must be flying away from it.

Sure. Why not?

From this, Lemaître developed the notion the the universe is expanding, and since that expansion all started when God said, 'let there be light,' then the stuff that exploded away fastest would be farthest away. Thus, Lemaître developed a theory that things in space were all flying away from Earth proportional to their distance.

The world of science now hinged on the assumptions of Copernicun gravity, flying photons, Cavendish Earth, gravity equality, 6.67, strange gases, random evolution, Darwin's timeline, neutral gravity, Helmholtz heat, constant light speed, Lorentz compression, flying electrons, nothing space, orbital electrons, Copernicun mass, floating continents, constant planet size, Eddington limitocity, omnipotent nothing, infinite photons, stellar fusion, Russian Expansion, doppelmatter, and that the universe sprang from a cosmic hatching.

In 1924, Louis Victor Pierre Raymond de Broglie, duc de Broglie (1892 – 1987) examined 'the wave nature of electrons' in his PhD thesis, going on to propose that matter also has wave-like properties. This concept, known as the Broglie hypothesis or *wave-particle duality* became a fundamental principle in the emerging field of quantum mechanics.

> *Physicists are still having the same debate that Aristotle was having one thousand, six hundred years ago. "Is light a wave or a particle?"*

Wave-particle duality did not resolve any paradoxes of the differences between waves and particles, but green-lighted the paradigm of ignoring it with the attitude of 'we'll figure that out later'.

Broglie also said,

> *"We proceed in this work from the assumption of the existence of a certain periodic phenomenon of a yet to be*

determined character, which is to be attributed to each and every isolated energy parcel".

In other words, "there is something really, really important that we're not catching on to." The **periodic** *phenomenon* Broglie is referring to is the phenomenon known as *time*. All planets obey it. All atoms obey it. All energies obey it. What is it?

Although general relativity boasted equations concerning time relative to motion and mass, it provided no intelligible concept of time nor any reason for time to exist.

The Copenhagen Interpretation

In 1927 the famous Bohr-Einstein Debates began during the fifth conference on quantum mechanics at Copenhagen, Denmark. The great issue plaguing both of Einstein's worlds – quantum mechanics and general relativity – were photon problems.

In relativity, photons were particles with mass flying around at the speed of light. Also in relativity, this was impossible. It was impossible for <u>any</u> object having mass to travel at the speed of light. Several problems would occur if something did.

- It would take infinite energy to achieve the speed of light, so every photon would have infinite energy.
- Momentum is proportional to mass, so if flying photons have *any* mass then they all have <u>infinite</u> mass.
- Then there's the problem of these infinitely massive objects impacting material objects billions of

times per second without having any infinite effect.

- Then you throw in anti-particle theory and you get these infinities crashing into those infinities, and all of that amounts to nothing.

In Copenhagen in 1927 Paul Dirac, Albert Einstein, and others were vigorously debating these problems when Dirac suggested that photons only exist when observed.

What did he mean?

Was that a sarcastic criticism of Einstein's theories?

There are two reasons why Dirac resorted to 'un-physics' to solve this physics problem.

On one side there were several experiments taken as evidencing the existence of photons as light: there were cathode rays, canal rays, anode rays, Compton effects, photosynthesis, and gravitational lensing.

On the other side was the double-slit experiment, the single-slit experiment, Arago's spot, cathode rays, canal rays, anode rays, Compton effects, gravitational lensing, and Heisenberg uncertainty. Those may all be taken to demonstrates that neither photons nor electrons exist in transmission.

> *"There is a doctrine well known to philosophers that the moon ceases to exist when no one is looking at it. I will not discuss the doctrine since I have not the least idea what is the meaning of the word existence when used in this connection. At any rate the science of astronomy has not been based on this spasmodic kind of moon. In the scientific world (which has to fulfill functions less vague than merely existing) there is a moon which appeared on the scene before the astronomer; it reflects sunlight when no one sees it; it has mass when no one*

is measuring the mass; it is distant 240,000 miles from the earth when no one is surveying the distance; and it will eclipse the sun in 1999 even if the human race has succeeded in killing itself off before that date."

— Arthur Eddington,

The Nature of the Physical World, 226

So how do photons exist at detection but do not exist in transmission?

Photons do not exist unless observed.

Actually, that is basically true.

As demonstrated in *Wave-Particle Transformation,* photons and electrons do not exist in transmission. It is important to understand, also, that light is never observed in transmission. Light is never observed unless received. When light is received by any object, that light is transformed through material interaction into photons.

But Dirac's use of the word 'observed' is wrong. Light-waves do not care whether or not some intelligent gray matter somewhere is paying attention. Neither do photons.

The resolution to the Copenhagen Interpretation, the Compton Effect, and all of the wave/particle paradoxes is the concept of *Wave-Particle Transformation* (WPT).

Perhaps when Dirac said that "Photons do not exist unless observed" he was pointing out that Einstein's flying photon theories could not be true.

Did Dirac realize wave-particle transformation? Probably not.

Neither he nor Einstein nor Bohr could figure it out. Not within their rule-book. This was the world of general relativity: Light consists of flying photons, space is nothing, and the speed of light is constant.

The assumptions of the world of Einstein made finding a solution to this problem impossible. The three fundamental paradoxes of Einstein – *The Nothing is Nothing Paradox, Universal Speed of Light Paradoxes,* and the *Flying Photon Paradoxes* – show that the very bases of the logic in question are false. With a false foundation of logic, no intelligent solution to this problem can be found, and none ever was.

Instead, we got the Copenhagen Interpretation.

The world of science now hinged on the assumptions of Copernicun gravity, flying photons, Cavendish Earth, gravity equality, 6.67, strange gases, random evolution, Darwin's timeline, neutral gravity, Helmholtz heat, constant light speed, Lorentz compression, flying electrons, nothing space, orbital electrons, Copernicun mass, floating continents, constant planet size, Eddington limitocity, omnipotent nothing, infinite photons, stellar fusion, Russian Expansion, doppelmatter, cosmic hatching, and none of this exists unless you look at it.

The Copenhagen Interpretation became hot sci-fi fodder, going on to be accepted as fact with 39% of scientists in a 2017 study supporting the Copenhagen interpretation. Study.

What about electrons? Those, too, only exist when observed. What about protons? Well, protons always have electrons so...

The Copenhagen Interpretation is the greatest paradox of the Age of Paradoxes. With this sci-fi/fantasy interpretation of reality there exists nothing which we have not laid eyes on. There exists nothing unless we have perceived it. There is nothing but our universe and we have created it through observation.

And to really get this party going, bring in some Heisenberg Uncertainty. When we Heisenberg, we destroy things when we observe them. The closer we observe an object, the less it exists.

THE UNIFIED THEORY

Bringing Einstein's two worlds together – quantum mechanics and general relativity – constitutes a unified theory.

Here is the new unified theory:

Copenhagen Heisenberg de Sitter Relativity

In this version of reality, the universe is empty and exponentially expanding. We, inside of this universe, perceive that it is full because we perceive. But when we observe closely, there is actually nothing there. In *Copenhagen Heisenberg de Sitter Relativity* we are altogether figments of each other's imaginations in an expanding universe composed of nothing.

This is the Age of Paradoxes.

The Bohr-Einstein debates continued formally and informally for decades. No resolution was every achieved.

Among the problems that persisted were photosynthesis. Photosynthesis traps light energy and re-emits that light energy later. If that was light, what is the speed of that light before it is re-emitted later?

To a lesser extent, why is the speed of light slower when it travels through material substances?

WAVE-PARTICLE-WAVE TRANSFORMATION DELAY (WPD)

Light waves travel at the speed of light. Photons do not travel.

When a light wave affects an atomic structure and induces a photon. That light-wave energy is stopped. That light wave energy now exists as a barely mobile subatomic particle.

A photon's mobility is comparable to the mobility of a balloon floating around in Earth's atmosphere.

The atomic submagnetic field, however, is constantly shifting. And when that field shifts a little bit, that photon is no longer neutral to the field. The photon then explodes – the aether-pi the photon was made of are scattered back into a gaseous state.

Submagnetic: Aether-pi interactions; proto-magnetic phenomena which do not involve whole atoms; proto-magnetic phenomena which affect individual sub-atomic units; atomic magnetic interaction.

Submagnetic phenomena include *gravity*, the *transmission of light*, the *strong force*, the *weak force*, the *electromagnetic effect*, the *transmission of magnetism*, induction and destruction of photons and electrons, quark-flipping, quark coloration, gluons, atomic bonds, and neutron-proton conversion.

The waves emitted when photons or electrons explode are light waves and they travel at the speed of light.

In the balloon allegory, that balloon exploded into light waves.

Wave-Particle-Wave Transformation Delay or Wave-Particle Delay (WPD) for short occurs when photons exist. Since photons do not travel, and since photons *contain the potential energy* of the light wave, the transmission of light is delayed whenever photons exist.

When photons explode, much of that wave energy travels only a very short distance *within the photon field* before inducing another photon. The shifting atomic field then destabilizes that new photon, and the process repeats until a photon is formed on the edge of the photon field.

Photon Field: The field of dense aether-pi surrounding an atomic structure within which photons may form. The photon field is larger than the electron field of the atom.

Electron Field: The field of highly dense aether-pi surrounding an atomic structure within which electrons may form. The electron field is smaller than the photon field of the atom.

Electrons are 'mega-photons' and obey all the same principles that photons do.

Okay, the balloon allegory is kind of weird at this point, but whatever. The exploded balloon's light waves cause another, virtually identical balloon to form elsewhere in the atmosphere. That process repeats until a balloon is formed on the edge of the atmosphere. When that last balloon explodes, the atmosphere beyond that point is too thin to support any balloon structure, so that energy leaves the Earth at the speed of light and is gone forever.

Wave-Particle-Wave Delay is what makes the speed of light slower when light is interacting with matter.

- The speed of light without material interaction is 299,792,458 meters per second.
- The speed of light in water is 225,650,000 meters per second.
- The speed of light in common glass is 191,863,000 meters per second.
- The speed of light in a gaseous plasma has been slowed to 3 meters per second.
- And with photosynthesis, the speed of light is reduced to zero meters per second.

WPD is why optical lenses bend light.

The rate at which WPD occurs varies proportional to frequency. The higher the frequency, the more rapidly wave-particle delay begins, so the faster that light is absorbed the more that light is lensed.

Since blue light is absorbed faster than red light, the speed of blue light in the prism is slower than red light. This lenses blue light inwards more than red light.

As the light continues through the prism, light is transmitted at a constant rate *proportional to frequency*. When light leaves the prism, it is lensed again, proportional to frequency.

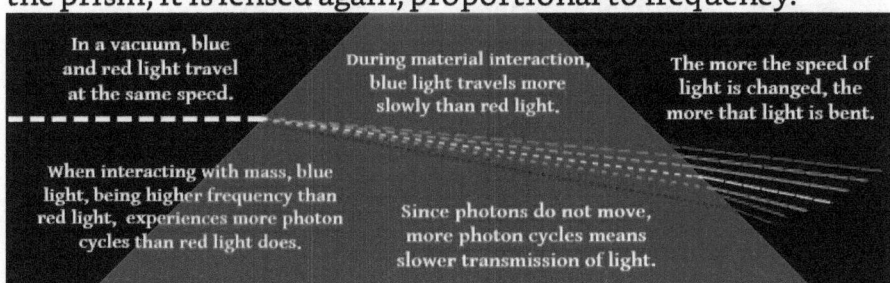

In a vacuum, blue and red light travel at the same speed.

During material interaction, blue light travels more slowly than red light.

The more the speed of light is changed, the more that light is bent.

When interacting with mass, blue light, being higher frequency than red light, experiences more photon cycles than red light does.

Since photons do not move, more photon cycles means slower transmission of light.

The reason for the additional delay, of blue light is that blue light has a higher frequency and therefore causes higher pressures in the aether-pi field. Since blue light causes more pressure than red light does, blue photons begin forming *farther away from the nucleus* than red photons do. The 'blue photon shell' of the atom is larger than the 'red photon shell' of the atom. Because of this, blue light experiences more wave-particle delay than red light does.

Blue light, being a higher frequency than red light, will also experience *more photon cycles* than red light. This also contributes to wave-particle delay.

Lensing happens on the surface. The 'surface' referred to is the collective photon field of the atoms of the prism. Theoretically, a single atom is capable of lensing light and forming an Arago Spot.

> To determine the direction of a concentric rainbow, look for the colors red and yellow. They will appear near to each other. The red is outside of the yellow. Red is the beginning of the rainbow.

> Red appears opposite of blue. Where red and blue seem to merge is where a new layer of the rainbow begins with blue.

> Consistent with gravitational lensing, a bright white halo exists within the rainbow rings of the Arago Spot in *Example 6* above.

At every material surface, gravitational lensing occurs. Light is turned inwards by gravitational lensing the same way that physical lens bends light. Three key things happen during the interaction.

Gravitational lensing, rainbow, and prism effect at border

Photons appearing at center prove light is not

traveling as straight-path photons.

Arago Spot by Aleksandr Berdnikov

- At the surface of the lens, light which affects that lens at an other-than-perpendicular angle becomes bent inwards proportional to its arrival angle,

- While that light is traveling through the lens, *wave-particle-wave transformation delay (WPD)* occurs when photons exist, and
- Light waves traveling in the dense aether-pi near atoms is transmitted more rapidly than light in a vacuum.

The net sum of these effects is that the speed of light is glass is 191,863,000 kilometers per second – 36% slower than light in a vacuum.

Gravitational Lensing

The first principle of gravitational lensing is that the speed of light is not constant, as shown below.

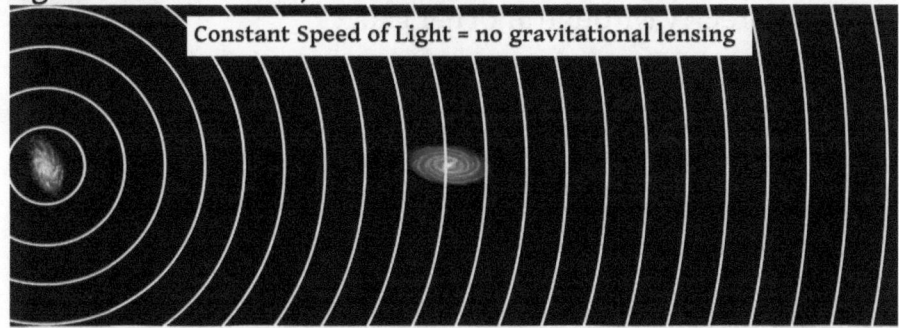

Constant Speed of Light = no gravitational lensing

Gravitational lensing involves all the same principles as ordinary physical lensing of light. The key difference, though, is light that is gravitationally lensed does not experience the wave-particle delay associated with glass and physical lenses. Because of this, gravity lenses are *acceleration lenses* whereas glass and water are *deceleration lenses.*

As shown below,

- As light enters the increasingly dense inertial frame of reference, light accelerates and becomes bent. When light is accelerated, like when light-waves leave water, the accelerated light-waves are lensed outwards.

- Acceleration continues to increase until the light passes its closest point to the galaxy.

- Light traveling within ten galactic radii experiences wave-particle delay proportional to its distance from the galaxy, that is. The close the light gets to objects, the more frequently it interacts with matter – gases – and gets slowed down, scrambled, and entropied, re-sulting in blind spot. It is very difficult to see a galaxy directly through another galaxy.

- Exiting the 'gravitational lens', the speed of light de-creases and light is turned inwards.

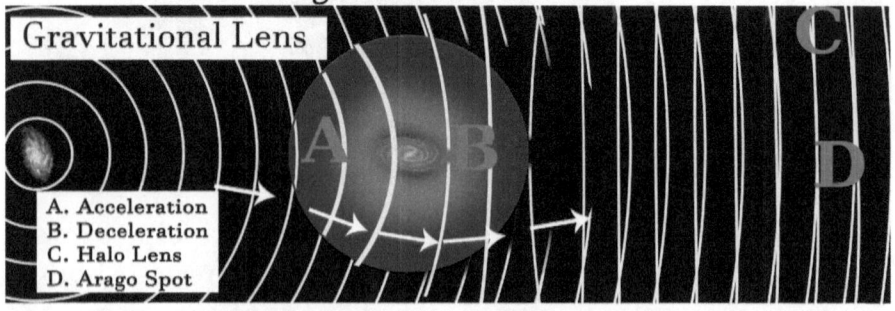

Gravitational Lens

A. Acceleration
B. Deceleration
C. Halo Lens
D. Arago Spot

One important result of lensing is interference. Since various portions of the same light waves are accelerated to different speeds, they end up recombined – at various points – at op-position where positive meets negative. That portion of that light wave energy in that region is constantly is destroyed through interferometry.

In 1927 Werner Karl Heisenberg (1901 – 1976) published his best known work – a work that laid the foundation of quantum mechanics – *The Heisenberg Uncertainty Principle*.

Heisenberg was the principal scientist in the Nazi Germany nuclear weapon project during WWII.

In 1928 Dirac published *The Principles of Quantum Mechanics*. A central concept uncovered was the Dirac equation – a way of describing spin ½ massless particles including electrons and quarks. This concept also introduced the prediction of *positrons* – subatomic particles opposite of electrons.

Unlike the electrons of Thomson and Rutherford, Dirac's electrons have no mass, and cannot, therefore, engage in any orbit.

Hubble's Red-shift

In 1929, Hubble's expanded, peer-reviewed report of galaxies was published including his concept of universal expansion and *Hubble's Law*.

Edwin Hubble's work was largely in field of spectroscopy. Using spectroscopy and applying the Dopplerian principles of waves, Hubble discovered a way to gauge the speeds of distant orbits.

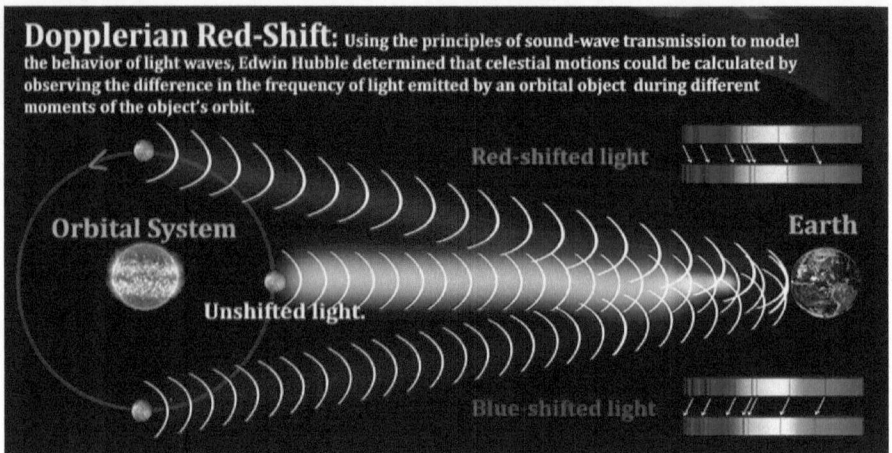

Dopplerian Red-Shift: Using the principles of sound-wave transmission to model the behavior of light waves, Edwin Hubble determined that celestial motions could be calculated by observing the difference in the frequency of light emitted by an orbital object during different moments of the object's orbit.

Orbital System

Red-shifted light

Unshifted light.

Earth

Blue-shifted light

Using the difference between red-shift and blue-shift, Hubble successfully measured speeds of faraway orbits. While this work was not original in concept, Hubble's equipment and accuracy shed new light on things.

And when Edwin Hubble identified distant galaxies, he noticed that the color of the light emitted by a galaxy seemed to correlate to the visual size of the galaxy.

Observing that little galaxies were usually redder and that bigger galaxies are usually bluer, and assuming all stars to emit same frequencies of light, Hubble concluded that the differences in the colors of the light he saw were dilation and compaction of white light.

That meant that white galaxies were basically stationary relative to Earth and that the redder galaxies were flying away.

As for the blue ones – the other 30% - which should be flying right at us, nobody either then or now believes that 30% of all the universe's galaxies are flying right at the Earth. Hubble did not believe it either, so he did not publish that part.

Currently – via Wein's Law – that blue-shift problem is explained as blue-shift due to high temperature. That is paradoxical in that that violates the basic premise that 'all galaxies are white'. The 'all galaxies are white' premise of Hubble's

Law implies that all galaxies are the same temperature.

So calculations of galaxy distances with Hubble's law are wildly erratic. With the ability to arbitrarily assign a temperature to a galaxy *and* arbitrarily assign a speed to that galaxy, scientists frequently generate controversial conclusions, including two famous 1998 reports. We'll get to that.

Hubble's Law

Hubble's belief in an expanding universe culminated in the creation of Hubble's Law:

Velocity = Hubble constant (a speed) * distance

$$v = Hd$$

Hubble estimated the Hubble constant to be 500 km/s/Mpc.

According to this law, the farther away anything is, the faster it is flying away from you.

Hubble's Law, unlike the law of Dopplerian red-shift, takes no account whatsoever of temperature, spectroscopic data, or anything else about a galaxy. It just says, 'the farther away it is, the faster is it going'.

According to this equation, the universe is expanding outwards from planet Earth.

The world of science now hinged on the assumptions of Copernicun gravity, flying photons, Cavendish Earth, gravity equality, 6.67, strange gases, random evolution, Darwin's timeline, neutral gravity, Helmholtz heat, constant light speed, Lorentz compression, flying electrons, nothing space, orbital electrons, Copernicun mass, floating continents, constant planet size, Eddington limitocity, omnipotent nothing, infinite photons, stellar fusion, Russian Expansion, doppelmatter, cosmic hatching, none of this exists, and the whole universe began at Earth.

Hubble's Law implicitly requires that the big bang happened right here on planet Earth in that objects cannot be traveling away from the Earth in every direction at speeds directly proportional to their distance *to Earth* unless it all started right here.

Hubble didn't believe in that part of it. Hubble expected to find a trend pointing to some other location. Hubble never believed, proponed, or supported any big bang theory. He just said the universe was expanding.

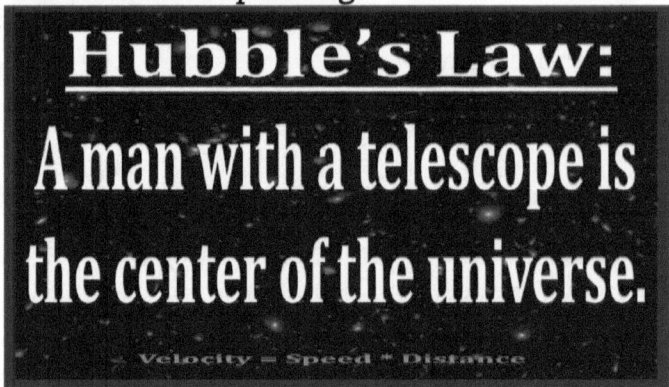

And that the galaxies are all flying away from us.

But they're not.

The closest major galaxy to our own is the Andromeda galaxy. That is not flying away from us. It is slowing lingering closer and closer. Of the next five major nearby galaxies, three are getting closer.

The closest galaxies to the Milky Way are not really getting any closer *or* father away. They are orbiting the Milky Way. They are satellite galaxies. There are at least of 48 of them and the largest one is called the Large Magellanic Cloud. None of those are obeying Hubble's Law.

Ignoring the blue problem, Hubble proposed that the universe was expanding. All distant galaxies were flying away from... somewhere. All 32 of them.

Hubble's universe was a very, very small place compared to modern catalogs of 400 trillion galaxies. That's about 399,999,999,999,968 more galaxies than Hubble was initially aware of.

Hubble's Law – which had no mathematical dependency on spectroscopy or red-shift – made it easy to theorize the distance of virtually any galaxy. That made it attractive and very popular.

But it does not work. It never worked.

Every thought a genius thinks is
not a genius thought.

In 1929 Fritz Zwicky (1898-1974), an American astronomer, rocket engineer, and jet engine engineer born in Belgium, pointed out that Hubble's red-shift/distance predictions had a margin of error far too large for that model to be feasible.

Zwicky posited, instead, that the observed red-shift was not due to relative motion. Instead, he proposed the theory of *Tired Light*.

Zwicky's tired light theory was that through some unidentified phenomenon light became tired and lost momentum during travel, and that that resulted in the observable red-shift.

Due to the errors and paradoxes of Hubble's theory of cosmo-logical red-shift, the theory of tired light has been put forward by many other researchers since then.

Tired light theory is paradoxical. The key to the tired light paradox is that light does not entropy. Light is entropy.

The concept of tired light is that the energy and frequency of the light has entropied.

How could the frequency entropy? How could the frequency become reduced? If I play a radio in the bathroom of my house, walk into the hallway, and close the door, the sound I then hear from that radio will be entropied. It will be harder to hear. If I go outside, it will be even harder to hear. If I go down the street, it will be very hard to hear. But wherever I go, the frequency of that sound will be same as the frequency heard in the bathroom.

Within flying photon theory, this idea is even worse.
-   How would a blue photon become a red photon?
-   Why would waves of light slow down where the speed of light is supposedly universally constant?
-   If we are getting red light which originated as blue light, then that light has lost a third of its energy and a third of its frequency.
    -   Where has that one third gone?
    -   Why don't we see that scattered as background light?
-   How can a photon, being a unit of energy, lose energy?

On March 18, 1929, Enrico Fermi was appointed a member of the Royal Academy of Italy by Mussolini, and he joined the Fascist Party on April 27th.

Ozone

In 1930, Sydney Chapman (1888 – 1970) realized how the ozone layer occurs.

The ozone layer is a layer of highly reactive oxygen compounds which transforms much of the ultraviolet radiation Earth gets from the Sun into infrared radiation, or heat. The ozone layer is thereby thought to protect the Earth from ultraviolet light and other high-energy radiation, including the gamma radiation from Eddington's fusion-based Sun.

The world of science now hinged on the assumptions of Copernicun gravity, flying photons, Cavendish Earth, gravity equality, 6.67, strange gases, random evolution, Darwin's timeline, neutral gravity, Helmholtz heat, constant light speed, Lorentz compression, flying electrons, nothing space, orbital electrons, Copernicun mass, floating continents, constant planet size, Eddington limitocity, omnipotent nothing, infinite photons, stellar fusion, Russian Expansion, doppelmatter, cosmic hatching, none of this exists, geocentric creation, and salvation by ozone.

Chapman's observations contradict the notion that the ozone layer protects the Earth. In fact, Chapman discovered that ultraviolet light supercharges atmospheric O_2 and that catalyzes a chemical reaction whereby the O_2s rearrange into O_3s. Atmospheric oxygen is transformed into ozone by ultraviolet light.

The ozone layer is not thought of as related to surface-level oxygen because the ozone layer occurs at a much higher altitude – fifteen to thirty-five kilometers above the surface. The reason that the ozone layer exists at such a high altitude is that ozone weighs 1/6th as much as atmospheric oxygen does.

In 1930, Wegener led a German expedition into Greenland. After using fossils to connect the whole world to Antarctica, Wegener was seeking fossils to see what he could attach Greenland to.

Ultimately, over-scheduled and under-supplied, Wegener and Rasmus Villumsen made a desperation move to escape the frozen wonderland: They took a dog sled team, no supplies, and headed east, eating the dogs one by one as they went.

Wegener was later found frozen stiff, buried by Villumsen. Villumsen was never seen again. Wegener's body is now beneath about 100 feet of snow somewhere in central Greenland.

In 1930 Paul Dirac Published *The Principles of Quantum Mechanics*.

Nucleogenesis

In an obscure 1931 paper, Albert Einstein posited that a universe which is constantly destroying matter is impossible. To overcome this problem, Einstein searched for a mechanism of nucleogenesis to counteract the destruction of matter.

In the Einsteiniverse, all stars destroy matter by converting it into energy. Also in the Einsteiniverse, black holes destroy both matter and energy.

The get the opposite of destruction of matter, Einstein predicted 'white holes' as counterparts to 'black holes' – white holes being objects which continually generate matter. Einstein attempted to mathematically connect the idea of white holes with his cosmological constant. Einstein abandoned that pursuit a short time later.

White hole theory – as a source of continued nucleogenesis to offset the effects of matter-destroying theoretical black holes – has been proposed several times since without mathematical or observational success.

But as Einstein observed, the universe would not be the way it is unless nucleogenesis was an ongoing process throughout the universe.

The world of science now hinged on the assumptions of Copernicun gravity, flying photons, Cavendish Earth, gravity equality, 6.67, strange gases, random evolution, Darwin's timeline, neutral gravity, Helmholtz heat, constant light speed, Lorentz compression, flying electrons, nothing space, orbital electrons, Copernicun mass, floating continents, constant planet size, Eddington limitocity, omnipotent nothing, infinite photons, stellar fusion, Russian Expansion, doppelmatter, cosmic hatching, none of this exists, geocentric creation, salvation by ozone, and that there must be white holes although they do not mathematically or observationally exist.

In 1932 Paul Dirac became the Lucasian Professor of Mathematics at Cambridge University. Dirac filled that position until 1969.

In 1932, Chadwick published *The Existence of a Neutron*. Chadwick proposed that the reason why we can't find four protons in a helium nucleus, but only two, is that the other two protons had no magnetic charge.

They obviously had mass. That was well-established. But the reason why we couldn't find them is that they are not positively (+) charged like protons are. They are neutral. They are neutrons.

What happens here is that when two or more protons are packed together, their combined *submagnetic* (proto-magnetic actions involving less than whole atoms) potential collapses an electron field onto the surface of a proton. The submagnetic field is localized, meaning that since one pole of the magnet is right on the surface of the other pole of the magnet, the is no exposed, external magnetic charge anymore. The submagnetic field is contained forming an *Atomic Faraday Cage*.

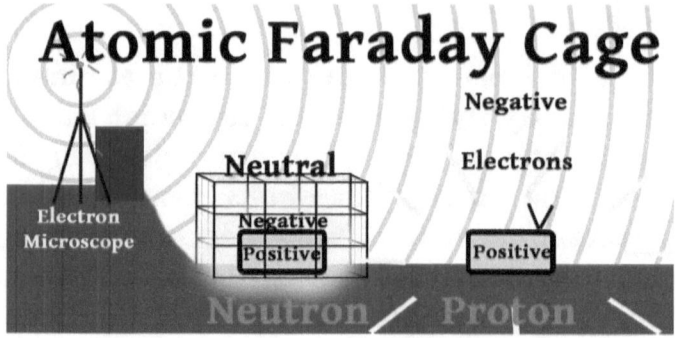

This phenomenon may be observed on planetary scales. Under gravitational pressure atmospheric gases may collapse into super-critical fluid state and completely change the electromagnetic and magnetic expression of a planet.

Gases in a super-critical state conduct electricity. Gases in a gas state do not. Because Earth does not have a supercritical atmosphere, the electrical inductions caused by movement of atmospheric gases builds up to hundreds of thousands of volts, producing aurorae and lightning storms.

Venus has 63 times as much atmosphere as Earth does, but Venus does not have lightning storms. When electrical potential occurs in the atmosphere of Venus, that potential is communicated to ground when voltage exceeds about five volts.

Earth has a powerful magnetic field extending 100,000 kilometers into outer space. Venus does not. Venus, until recently, was believed to have no magnetic field at all. Venus may have a very powerful magnetic field, but it barely reaches beyond Venus' atmosphere.

Earth is like a proton – a big magnetic field with lightning and electron storms all over the place. Venus is magnetically and electrically self-contained like a neutron.

If Venus lost 80% of its gravity, it would lose its atmosphere, lose its electrical protection, cool off, and experience lightening storms.

If the neutron becomes divided from the proton, the electron field will 'peel away' in about 14.65 minutes, at which point in time, the neutron has become a hydrogen atom, being a proton with an electron field.

1932
Nobel Prize
in Physics

Werner Karl Heisenberg

"for the creation of quantum mechanics, the application of which has, inter alia, led to the discovery of the allotropic forms of hydrogen."

The discovery of the neutron demonstrated how difficult it might be to really understand what nature is made of.

Origin of the Universe

In 1932, Willem de Sitter coauthored a paper with Albert Einstein which advanced the concept of an empty universe exponentially expanding. This hypothetical and purely theoretical situation is called a *de Sitter universe* which is filled with *de Sitter space.*

1932
Nobel Prize
in Physics

Werner Karl Heisenberg

"for the creation of quantum mechanics, the application of which has, inter alia, led to the discovery of the allotropic forms of hydrogen."

The de Sitter universe is an advancement of big bang theory. De Sitter universe is the theory used to famously "calculate the age of the universe to within 10-33 seconds" after a big bang.

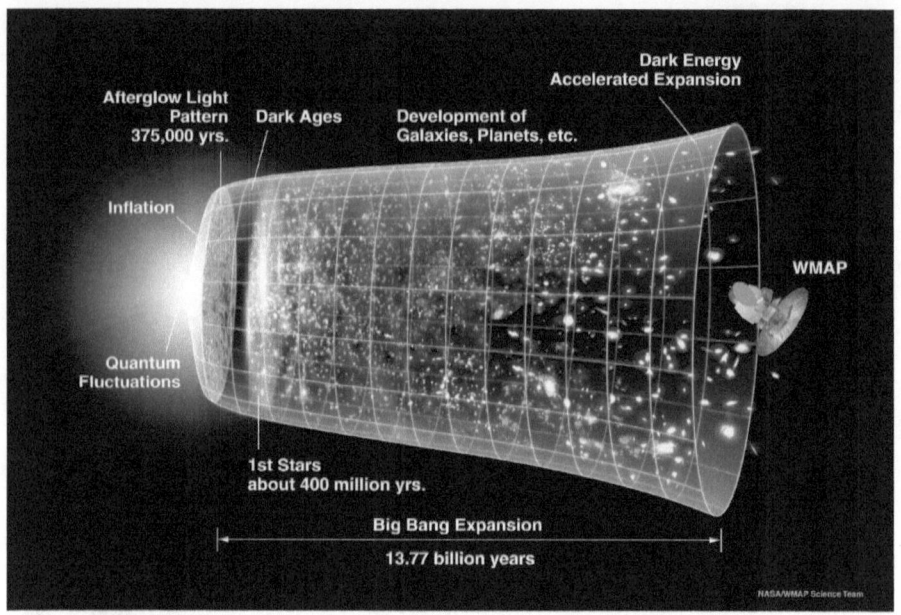

In big bang theory, after the cosmic egg explodes super-dense energy spreads. Nucleogenesis occurs later. De Sitter space is thought to exist during that period of time before nucleogenesis.

In the NASA WMAP picture above, a de Sitter universe would have existed during the *Dark Ages.*

The concept of a de Sitter universe has subsequently become attached to *dark energy* and/or *dark matter* theory which both began a short time later.

In 1932, the first cyclotron was patented out of the University of California at Berkeley in the USA.

Invented by Ernest Orlando Lawrence (1901 – 1958), the cyclotron was first tested in 1929-30. The cyclotron was the first particle accelerator.

Taking Aston's mass-spectroscopy up a few thousand volts, Lawrence took the art of electromagnetic manipulation of matter up from tearing chemicals apart up to smashing atoms together.

Now we could start to see what they're really made of.

Lawrence smashed electrons. He couldn't find any mass. That was not a surprise because the supposed masses were so small he might not detect them anyway.

Then Lawrence started smashing atoms together. Contrary to Einstein's predictions, the did not explode. Most of them, he determined, were taking energy in instead of putting energy out when he destroyed them.

Also in 1932, Einstein left socialist Nazi Germany for America, where he took a position at Princeton University in New Jersey.

In 1933 Leo Szilard conceived of the idea of a *nuclear chain reaction* – the basis for an *atomic bomb.*

In 1934, Szilard and Enrico Fermi patented the idea of a nuclear reactor.

Dark Matter

In 1933, Fritz Zwicky viewed the Coma Galaxy Cluster. From these observations, it was determined that there must be something more, something else... something besides gravity keeping them together..

1933
Nobel Prize
in Physics

**Erwin Schrödinger and
Paul Adrien Maurice**
"for the discovery of new
productive forms of atomic theory."

Based on Zwicky's calculations of Coma Cluster objects' average kinetic energy, Zwicky calculated that there was *dunkle materie* (dark matter) there which totaled 400 times the mass of the observable objects. Otherwise, the velocities of the various galaxies should be pulling them apart.

Zwicky described dark matter as an extra-galactic substance holding the cluster together. It was dunkle (dark) in that it had no visible signature. It was materie (material) in that it had an apparently material attribute – gravity.

The world of science now hinged on the assumptions of Copernicun gravity, flying photons, Cavendish Earth, gravity equality, 6.67, strange gases, random evolution, Darwin's timeline, neutral gravity, Helmholtz heat, constant light speed, Lorentz compression, flying electrons, nothing space, orbital electrons, Copernicun mass, floating continents, constant planet size, Eddington limitocity, omnipotent nothing, infinite photons, stellar fusion, Russian Expansion, doppelmatter, cosmic hatching, none of this exists, geocentric creation, salvation by ozone, white holes, and that dark matter controls the structure of the universe.

In 1934, Fritz Zwicky and his colleague Wilhelm Heinrich Walter Baade (1893 – 1960) – a German American astronomer – coined the term 'supernova', which Zwicky was ardently studying. Zwicky independently identified 120 supernovae – more than anyone else ever did without a computer.

All Tied Up - The EPR Paradox

In 1935, Albert Einstein, Podolsky, and Rosen (EPR) found a paradox within general relativity. Relativity's law predicted instantaneous action at a distance which Einstein called 'spooky action'. Spooky action occurs if any information, action, or motion occurs at a speed greater than the speed of light.

In the Einstein Michelson Morley Lorentz universe nothing could transfer or occur faster than the speed of light, so the spooky action predicted through Einstein's work *violated* Einsteins own work.

1935
Nobel Prize
in Physics

James Chadwick
"for the discovery of the neutron."

That is the definition of a paradox.

Instantaneous action implies infinite speed. Infinite speed implies infinite force. Einstein did not like theses infinities because, unlike his other infinities – gravity from nothing, for instance – these infinities did not support his theory, but contradicted it.

The key to Einstein's conundrum was gravity in Einstein's empty space. If space is empty, then gravity must be conveyed or delivered through particle interaction. According to Einstein's rules, nothing could transmit faster than the speed of light.

The problem here emerges with orbits. It is conventionally thought that it takes light eight minutes to reach Earth from the Sun, and if gravity is transmitted as particles, then it takes the same eight minutes for gravity to reach the Earth.

If gravity arrives at Earth eight minutes after it leaves the Sun, where will that gravity encounter Earth? It would be like shining a beam of light at the Earth. Where would the light arrive?

It wouldn't. That light and gravity would arrive after Earth's position had changed 14,860 kilometers - 117% Earth's radius. That light and gravity would completely miss Earth by 2,170 kilometers.

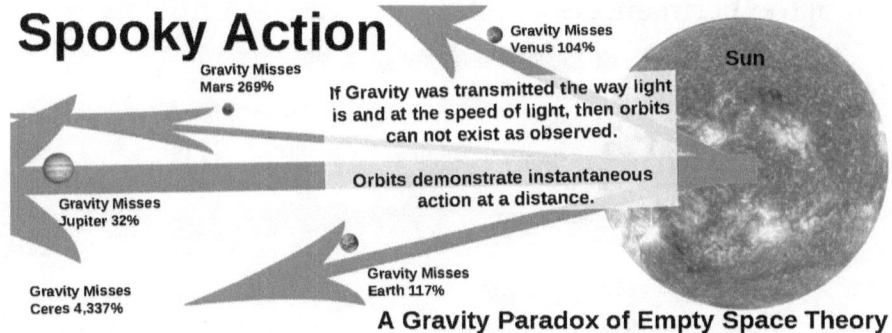

Spooky Action

Gravity Misses Venus 104%

Sun

Gravity Misses Mars 269%

If Gravity was transmitted the way light is and at the speed of light, then orbits can not exist as observed.

Orbits demonstrate instantaneous action at a distance.

Gravity Misses Jupiter 32%

Gravity Misses Ceres 4,337%

Gravity Misses Earth 117%

A Gravity Paradox of Empty Space Theory

The EPR Paradox is that if action such as gravity happens instantaneously, then the relativistic principles concerning the speed of light are false or conditional.

The EPR Paradox is Einstein's most referenced work.

The world of science now hinged on the assumptions of Copernicun gravity, flying photons, Cavendish Earth, gravity equality, 6.67, strange gases, random evolution, Darwin's timeline, neutral gravity, Helmholtz heat, constant light speed, Lorentz compression, flying electrons, nothing space, orbital electrons, Copernicun mass, floating continents, constant planet size, Eddington limitocity, omnipotent nothing, infinite photons, stellar fusion, Russian Expansion, doppelmatter, cosmic hatching, none of this exists, geocentric creation, salvation by ozone, white holes, dark matter, and the acceptance that these laws clearly contradict themselves.

The principle of spooky action was countered by Bohr with the concept called *quantum entanglement.* Quantum entanglement holds that instead of the two particles having a magical permanent connection, those particles are, instead, not connected until they are observed.

Quantum entanglement is an extension of the Copenhagen interpretation where photons do not exist unless observed.

Quantum entanglement has been reported as being evidenced through experiments resembling the illustration below

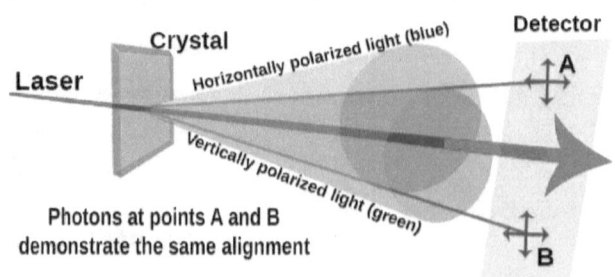

Experiment Posited as Evidence of Quantum Entanglement

The principle of quantum entanglement is that, in the world of Einstein's flying photons, the photons which flew from the crystal to the receiver flew through empty space and, based on Heisenberg uncertainty, should have assumed spins independent of each other.

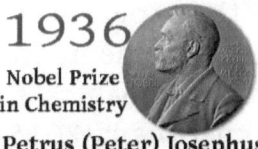

1936
Nobel Prize in Chemistry

Petrus (Peter) Josephus Wilhelmus Debye
"for his contributions to our knowledge of molecular structure through his investigations on dipole moments and on the diffraction of X-rays and electrons in gases."

Advanced Double-slit Experiment

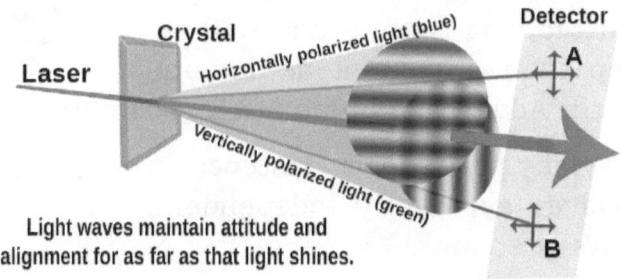

Laser
Crystal
Horizontally polarized light (blue)
Vertically polarized light (green)
Detector
A
B

Light waves maintain attitude and alignment for as far as that light shines.

When seen as light of waves instead of particles, the experiment proposed as a proof of quantum entanglement serves, instead, as an advanced double-slit experiment, as shown in Illustration Advanced Double-Slit experiment.

1936
Nobel Prize in Physics

Victor Franz Hess
"for his discovery of cosmic radiation"
Carl David Anderson
"for his discovery of the positron."

If light did not travel as waves then the photons which occur at points A and B should not align to the original signal. If light does travel as waves, then those waves exist for as far as that light is transmitted and 'entanglement' is impossible to avoid.

Contrary to popular misconception, Einstein's spooky action is not the same or even similar to Bohr's quantum entanglement. Einstein's space-time was made of nothing, so gravity was not a particle event.

1937
Nobel Prize in Chemistry

Walter Norman Haworth
"for his investigations on carbohydrates and vitamin C"
Paul Karrer
"for his investigations on carotenoids, flavins and vitamins A and B2."

If the experiment above is taken as proof of Bohr's quantum entanglement, then quantum entanglement violates quantum mechanics. A signal divided by 143 kilometers in a later experiment (2015) ought to be significantly disturbed by Heisenberg uncertainty.

If quantum entanglement works, then Heisenberg uncertainty does not.

But if light is transmitted as waves then those will be waves for as far as that light is transmitted, and those waves will always keep their alignment as demonstrated in the Advanced Double-Slit Experiment.

1937
Nobel Prize
in Physics

Clinton Joseph Davisson
and George Paget
"for their experimental discovery of the
diffraction of electrons by crystals."

In 1937 Zwicky posited the existence of 'gravity waves' apparently independent of Einstein's assertions.

1938
Nobel Prize
in Chemistry

Richard Kuhn
"for his work on
carotenoids and vitamins."

On December 22, 1938, Hahn and Strassmann sent a manuscript to Naturwissenschaften which reported their discovery of *barium*. They had generated barium by bombarding uranium nuclei with neutrons. They reasoned that they had *divided* or fissioned the uranium nucleus into the smaller, newly discovered element.

1938
Nobel Prize
in Physics

Enrico Fermi
"for his demonstrations of the existence of
new radioactive elements produced by
neutron irradiation, and for his related
discovery of nuclear reactions brought
about by slow neutrons."

LuftHeinkel

Jet engines were proposed as early as 134 CE. The first commonly known instance of a working jet engine was one built in 1791 by an Englishman named John Barber (1734–1793) and used to fly a short distance attached to a large kite.

The Wright Brothers – Wilbur (1867 – 1912) and Orville (1871

– 1948) – flew the first airplane (with an automotive engine), and did so in North Carolina on December 17, 1903.

That was all well and good, but until jet engines and airplanes were combined into high-speed physics, the mysterious force of gravity wasn't going to give up any secrets.

1939
Nobel Prize in Chemistry

Adolf Friedrich Johann Butenandt
"for his work on sex hormones"
Leopold Ruzicka
"for his work on polymethylenes and higher terpenes."

In 1939, the Heinkel He 178 became the first airplane powered by a turbojet.

In 1939 Otto Robert Frisch (1904 – 1979) born in Vienna, Austria-Hungary, was working in Germany and confirmed the findings of splitting uranium by Hahn and Strassmann. The German government began investigating atomic bombs.

In 1939 Bohr and Wheeler published a paper titled *The Mechanism of Nuclear Fission.*

In 1939, Szilard contacted his old friend, Albert Einstein, a man of fame and prestige in American academic circles, and explained the German nuclear situation, asking for Einstein's help.

On August 2, 1939, Einstein wrote a now-famous letter to President Franklin D. Roosevelt, explaining the weapons potential of a nuclear bomb. In response to Einstein's letter, President Roosevelt established the Advisory Committee on Uranium in October 1939. Quickly establishing laboratories in New York, one project that developed was called the *Manhattan Project.*

World War II began on September 1, 1939 and lasted six years.

The Nobel prizes were halted, in part, because giving money to scientists in times of war when those scientists are designing weapons *for* war can be seen as an *act of war* favoring the recipient's country.

No Nobel Prizes Awarded
1939 - 1944 No Nobel Banquet was held.

Further, inventing weapons for war is, arguably, not 'for the greatest benefit for mankind'.

In 1940 John Archibald Wheeler (1911 – 2008), an American theoretical physicist called Richard Feynman. Feynman related, "I received a telephone call one day at the graduate college at Princeton from Professor Wheeler, in which he said, "Feynman, I know why all electrons have the same charge and the same mass!"

Wheeler speculated that all electrons and their counterparts – positrons – were one, individual, singular object. At the time, matter was thought to consist of electrons and positrons, so the *one electron universe* model posited that all matter was the experience of a single electron bouncing back and forth in time.

Wheeler also advanced on the *Copenhagen Interpretation* with the doctrine of *it for bit*. Near the end of his life, Wheeler summarized it-for-bit thus,

> *"It from bit symbolizes the idea that every item of the physical world has at bottom — at a very deep bottom, in most instances — an immaterial source and explanation; that what we call reality arises in the last analysis from the posing of yes-no questions and the registering of equipment-evoked responses; in short, that all things physical are information-theoretic in origin and this is a participatory universe."*

A big step up from the Copenhagen interpretation, the it-for-bit universe not only fails to exist unless observed, it must *choose* whether or not to *participate* in being observed.

Wheeler is also famous for coining the term *wormhole, black hole,* and *neutron moderator (nuclear fission physics).*

In 1941 Andrew McKellar (1910 – 1960) discovered what

would later come to be known as *cosmic microwave background radiation.* McKeller was a Canadian astronomer whose parents immigrated from Scotland. He went to America to attend the University of California where, in 1941, he was studying the idea of the presence of the existence of cyanogen (CN) and methyne (CH) in interstellar space when he noticed the ambient radiation.

Expecting to find CN and CH emissions as the source, McKeller was surprised to determine that there was no apparent source, and spectroscopic analysis of the CMB did not indicate a chemical source. Like light emitted by matter when very excited, the CMB was a continuous spectrum.

In December of 1941, Edwin Hubble reported to the American Association for the Advancement of Science that results from a six-year survey with the Mt. Wilson telescope did not support the theory that the universe was expanding.

The catalog of galaxies had grown from 32 to well over 5,000 by that point in time. Hubble, based on his surveys, was greatly dissatisfied with his former concept of universal expansion and disgusted with the idea of big bang theory.

Hubble stated,

> *"The nebulae (galaxies) could not be uniformly distributed, as the telescope shows they are, and still fit the explosion idea. Explanations which try to get around what the great telescope sees fail to stand up. The explosion, for example, would have had to start long after the earth was created, and possibly even after the first life appeared here."*

For better or worse, Darwin's evolution theory was still in charge of creation.

Going Nuclear

In 1942, Julius Robert Oppenheimer took the lead of President Roosevelt's new *Development of Substitute Materials* program – a programs later known as *The Manhattan Project.*

On December 2, 1942, at the University of Chicago, Illinois, the first nuclear reactor – Chicago Pile 1 – went critical. The Manhattan Project had achieved its first objective – proving that a chain fission reaction was possible.

1943
Nobel Prize in Physics

Otto Stern
"for his contribution to the development of the molecular ray method and his discovery of the magnetic moment of the proton."

In July of 1944, Germany and England both deployed jet-powered fighter airplanes against each other. They were generally ineffective and consumed five times as much fuel as standard propeller-driven aircraft. Those clumsy little jet-airplanes hadn't learned the lessons that would be learned at Los Alamos.

1945
Nobel Prize in Physics

Wolfgang Pauli
"for the discovery of the Exclusion Principle, also called the Pauli Principle."

In 1945 Germany surrendered.

On July 16, 1945, in the desert of New Mexico, *The Gadget* sat on the ground. It was about to be detonated. The project was called *Trinity.* At 5:30am the Gadget became the first atomic weapon ever detonated.

Following the success of Trinity, nuclear technology was swiftly applied to crushing the Japanese. On August 6, 1945, Hiroshima and Nagasaki were bombed.

That, coupled with Japan's long string of defeats in the Pacific, led to the Japanese Emperor's unconditional surrender to the United States of America, formalized on September 2, 1945.

Heroshima & Nagasaki

NEW NEWTONIAN JET PHYSICS

The USA, which seemed 'late to the game' of jet-airplane technology, had been developing technology primarily under the shroud of military secrecy. This had also been true of Germany and England.

And scientists in the USA, as Americans so often do, were demanding more. And when they put more power into their engines, they found them increasingly prone to crashing. Something was wrong.

Gravity was wrong.

When engineers designed their jet-aircraft, they designed them to fly in such a manner as resists gravity pulling downwards at 90°. This was the angle of gravity as calculated via Newton's Law of Gravity and the Cavendish model of Earth. Unadjusted for the surface stood upon, those models counted gravity as occurring from a central point at the middle of Earth.

Accounting for wind-drag, scientists designed aircraft to

maintain a thrust angle of ~89 ° while cruising and and as low as 79 ° while taking off.

Test pilots like Chuck Yeager reported that the jet planes operated better during take-off than in flight. Landings were the worst.

When reliance on standard theory didn't work, a few brave engineers (two with experience as test pilots) calculated that the way that pilots were holding their flaps during flight meant that the airplanes were being flown at a much lower thrust angle than predicted. Applying this, engineers moved the wings forward to decrease the thrust angle.

At 75 °, planes worked better. At 60 °, a lot better. At 50 ° they flew great. At 40 ° they did not fly. The sweet spot was near 45 ° as shown *Aeronautics Thrust Angle*

This was and is the evidence that Newton and Cavendish never had. This is the angle of gravity. The thrust angle *is* the angle of gravity.

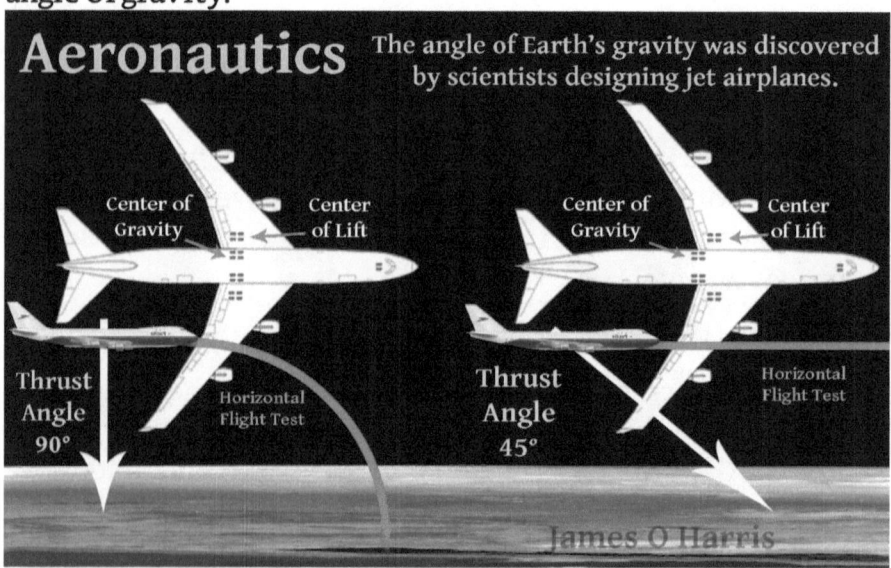

Other countries continued struggling against the realities of gravity without getting to heart of the problem. These problems did not only plague their profitability, these oversights made their products inferior.

British engineers, for example, shifted the gravitational center of the airplane towards the rear to counteract the unpredicted angle of gravity.

The first commercial jet airplane, the Havilland DH-106 Comet, debuted in 1952 with a long nose and a heavy tail.

Havilland DH-106 Comet

These modifications placed huge amounts of stress of the fuselage (the main body of the aircraft). This, combined with two other problems, took the Havilland Comet down in flames.

The two other problems were, one, the windows on the airplane were square. Really the squareness of the windows was not the problem. The problem was that the internal structure of the fuselage was likewise square in its interconnectedness. Structurally speaking, squares are very weak.

Within about one year of their debut, three Comets had experienced catastrophic fuselage failures while in flight.

The other fuselage problem was the error of not *over-designing* the fuselage to account for unexpected agitation. 'Unexpected agitation' of the fuselage occurs when pilots either take off or land with the nose of the aircraft too high in the air. Being too close to the ground at too steep of an angle, the tail of the aircraft strikes the runway. This can occur due to pilot error, use of under-sized airstrips, or unexpected headwinds. Tail-dragging most commonly occurs during landing.

A tail-dragging landing is typically successful. The structural damage due to that landing may not become apparent until the plane attempts to take off again or during subsequent operation. This is something to watch for when choosing airlines, airplanes, or pilots – look for denting and/or drag marks on the underside of the airplane between the wings and the tail. If there is any there, don't fly.

The Havilland Comets, having heavy tails and weak fuselage, were highly susceptible to critical injury in consequence of unexpected agitation.

In 1953, about one year after its debut, a Havilland Comet crashed after takeoff from Calcutta. On January 10, 1954, another one crashed, this time after taking off from Rome. The fleet was then grounded.

Britain was heavily invested in the Havilland as their economic salvation following World War II. The British government fast-tracked re-approval and the Havillands took off again on April 1, 1954.

On April 9, 1954, another one crashed, this time killing everyone on-board.

The Boeing 707 – with vastly superior engineering – was tested in 1954 and debuted in 1958. The 707 featured smaller wings pushed forward.

Boeing 707

This design shifted more of the turning torque to the rear of the craft, so the 707 also featured a much larger tail fin (*vertical stabilizer*).

Other American aircraft such as the McDonnell Douglas DC-8/10 and Lockheed's L1011 Tristar were also shaped like the Boeing 707.

Lockheed Tristar L1011

The Boeing 707 became the international standard for large aircraft design.

McDonnell Douglas DC-8

In 1949 the USSR detonated the first hydrogen bomb – an atomic bomb with a lot more punch. This began the arms race between America and the USSR, ultimately initiating *The Cold War.*

The cold war was unique in that it involved both countries

spending as much on their militaries as they did in wartime, but without actually fighting each other.

1950 - 2000

In January of 1950, de Havesy precipitated the metal of the Nobel medals from the aqua regia and returned it to the Nobel Foundation. The medals were carefully restruck and officially re-awarded in 1952.

In 1955 John Wheeler invented the concept of *quantum foam.*

In the quest to resolve EPR and other paradoxes, Wheeler suggested that the Heisenberg uncertainty principle could be caused by quantum foam, and that the "very geometry of space-time fluctuates" due to this foam. Fluctuation in this quantum foam were speculated to cause *vacuum pressure* with a non-zero energy known as *vacuum energy*.

This entity occupying empty space is another theory of aether within quantum mechanical theory.

A Smaller Hubble

In 1956, Baade examined many *cepheid variable* stars and, comparing them with former records, recalculated Hubble's constant to 65km/MPc/s – merely 13% of the speed Edwin Hubble had calculated.

Cepheid stars are characterized by emitting pulsations of light at regular intervals and appearing to increase in diameter.

Cepheid variable stars most likely consist of two objects in close, binary orbit.

Baade's new Hubble constant made Hubble's constant more

believable. Maybe.

The new, lower Hubble constant spawned a new problem. Now, instead of inter-galactic dark matter holding things together the way Zwicky had estimated, an opposite force became necessary – dark energy. Now something had to be holding them apart or they *should* all be crashing together.

That conversation wouldn't really hit the front page until 1998.

In 1957 Hugh Everett (1930 – 1982) proposed the *many worlds interpretation* of quantum mechanics. Many worlds is the idea that every decision which occurs incurs a new and unique universe. Of course, only one universe is observed, so this logic is attached to the Copenhagen interpretation where existence is optional. This logic, like the Copenhagen interpretation and the EPR paradox, spawned from the paradoxes of wave-particle duality.

Quantum Gravity

Peter Ware Higgs (1929 –) is a British theoretical physicist and emeritus professor in the University of Edinburgh who, in 1964 proposed the existence of the *Higgs Particle.* Higgs was not necessarily the foremost contributor to the theory, but he had the best name, the team decided, so they named it after him. Other contributors included Kibble, Guralnik, Hagen, Englert, and Brout. The Higgs boson theory emerged from the 1964 PRL symmetry breaking papers by these gentlemen.

The *Higgs mechanism* postulates the existence of the *Higgs field* and that that field confers mass on quarks, leptons, and other *fermions.* Higgs' later papers proposed a *Higgs boson* because, "How could the field do anything if it wasn't made of anything?"

So Higgs populated the field with Higgs' bosons.

The reasoning for the necessity of the Higgs boson was that for any object to gravitationally affect any other object, there had to imbalances – spontaneous symmetry breaking – they said.

The logic works something like this: If a proton had mass, and that mass was inside, and that mass was proportional to gravity, then what is *outside of the atom* which conducts, creates, or transmits gravity? There must exist some tangible connection or gravity would not happen.

Higgs et all proposed that this meant that *portions of an atom's mass* must be momentarily transferred through an external mechanism and exchanged with another atom in order for the force of gravity to exist.

The mechanism proposed for that transfer was the Higg's boson. It was a boson in that it is a perpetual, persistent entity and it may have or evince the attribute of mass.

The Higgs boson theory was the next great hope of quantum mechanics. Quantum mechanics has never succeeded in quantifying gravity. Without redefining gravity, the world of quantum mechanics can never satisfy the expectation that it will become a unified, universal, or complete theory.

The mysterious Higgs boson – the secretive and unfindable holder of all mass – is seen as the 'final' proof needed to justify general relativity, the key to a unified theory. As the holy grail of physics, it also called *the God particle.*

As with other saviors, its arrival seems a bit delayed.

The Higgs mechanism would be valiantly experimented towards for nearly fifty years before anyone generated enough of a part of an evidence to be able to nearly affirm the possible existence of the Higgs mechanism.

As with other saviors, some say he never showed up.

As with with Cavendish Earth, though, no one in the quantum mechanics world had come up with anything better, so the

Higgs mechanism theory was generally accepted as true.

With 'nothing space' now densely populated with mass-giving Higgs bosons, and the quantum mechanics community largely accepting this idea, quantum mechanics became further steeped in aether theory.

The first most obvious problem with that is that it ruins flying photon theory – the basis of quantum mechanics. If space if full of Higgs particles, then flying photons would have to interact with that field, blurring light from distant galaxies and resulting in Heisenberg uncertainty.

Quantum mechanics unwittingly gave gravity to aether theory, but if they're going to get things right, they will have to do a whole lot more than that. QM has over 30 unique 'fields' corresponding to diverse *extra-atomic* phenomena. Every one of these 'fields' is an aether theory in its own right.

The world of science now hinged on the assumptions of Copernicun gravity, flying photons, Cavendish Earth, gravity equality, 6.67, strange gases, random evolution, Darwin's timeline, neutral gravity, Helmholtz heat, constant light speed, Lorentz compression, flying electrons, nothing space, orbital electrons, Copernicun mass, floating continents, constant planet size, Eddington limitocity, omnipotent nothing, infinite photons, stellar fusion, Russian Expansion, doppelmatter, cosmic hatching, none of this exists, geocentric creation, salvation by ozone, white holes, dark matter, EPR, and Higgs boson.

The most obvious resolution to this conundrum is that all field effects are exactly that – events within a field. This field is composed of a substance which has an affinity for and/or interactive relationship with matter and appears indivisible from matter. Therefore this field is taken to be an *attribute* of matter

Seeing as this field has real effects and real attributes in its own right, it may also be reasoned that matter is an attribute of the field and not the other way around. Light is transmitted where no matter exists. The field exists with or without matter.

The 'field' and all its effects are, collectively, *the Inertial Frame of Reference* and we'll get into that more later.

Cosmic Microwave Background Radiation

In 1964 American astronomers Arno Allan Penzias (1933 -) and Robert Woodrow Wilson (1936 -) discovered the *cosmic microwave background radiation (CMBR OR CMB)* with the use of a 6-meter horn telescope.

It wasn't so much that they *discovered it* as *couldn't get rid of it.* Trying for clean, clear pictures at low frequencies, Wilson and Penzias did all they could to eliminate this unknown noise from their surveys. They tried it day and night, at different angles, at different times of year. They climbed up there, evicted the pigeons, and removed the nests. Nothing stopped the signal.

Without a basis for what was causing it, CMB was reported.

Robert Henry Dicke (1916 – 1997) was an American physicist of Princeton and MIT who proposed that the CMB was a remnant of an explosion indicating a big bang. Dicke oversaw the doctorate of Phillip James Edwin Peebles (1935 –) who is also credited with theorizing the existence of primordial CMB from a big bang.

The problem with that theory is that for ultraviolet light (light associated with an explosion) to be Dopplerian red-shifted into CMB frequencies, that radiation must be traveling at thirty times the speed of light away from the Earth.

Conventional reports go for a much more powerful bang, that, instead of UV light, starts with gamma radiation instead. Gamma radiation has a frequency ten million times higher than ultraviolet radiation, so that model requires a big bang explosion traveling outwards at thirty million times the speed of light.

If light is traveling away at more than one (1) speed of light, then no light would arrive on Earth.

Satellites were later sent up to map the CMB.

The world of science now hinged on the assumptions of Copernicun gravity, flying photons, Cavendish Earth, gravity equality, 6.67, strange gases, random evolution, Darwin's timeline, neutral gravity, Helmholtz heat, constant light speed, Lorentz compression, flying electrons, nothing space, orbital electrons, Copernicun mass, floating continents, constant planet size, Eddington limitocity, omnipotent nothing, infinite photons, stellar fusion, Russian Expansion, doppelmatter, cosmic hatching, none of this exists, geocentric creation, salvation by ozone, white holes, dark matter, EPR, Higgs boson, and that we see microwaves which are light traveling away as gamma rays at thirty million times the speed of light.

In 1974 Kaku Michio (1947 –) and Professor Keiji Kikkawa (1935 – 2013) of Osaka University in Japan brought string theory to life. String theory is something like the vibrating strings of a million million orchestras but the stings aren't attached to anything and form the aether and reference frame and gravity and everything.

1974
Nobel Prize in Physics

Sir Martin Ryle and Antony Hewish

"for their pioneering research in radio astrophysics: Ryle for his observations and inventions, in particular of the aperture synthesis technique, and Hewish for his decisive role in the discovery of pulsars."

This modification to quantum mechanics appeared as a reasonable alternative to the downfallen quantum foam idea.

String theory wasn't really something new, but Michio and Kikkawa's paper described string theory in field form and suggested it as the underlying field or aether theory for quantum mechanics.

This quantum aether theory called 'string theory' claims to encompass all the 30-plus fields of quantum mechanics into a single field. A good idea, for sure, but string theory makes no testable predictions.

There's no way to prove it right, there's no way to prove it wrong.

Strategically, that's a great way to generate endless debates, naval staring, and never-ending stories. If you don't claim to be right about anything, how can anyone prove you wrong? What are you claiming to be right about?

String theory also predicts other universes and allows for a multiverse where everything happens and doesn't happen, depending on your choice of universe.

The theory of a multiverse inherently dictates that multiple universes cannot be simultaneously or consecutively visited so, once again, there is no testable basis for this theory.

The most obvious problem here is that *matter is*. If there were billions of other dimensions consecutive to our own, where would they fit? The maximum compaction of atomic nuclei is mathematically defined as the density of a neutron star. That would be 4E+14 g/cm3 or 1.2E+14 times denser than matter on average. That's an absolute maximum local space limit of 1.2E+14 universes – 1.2E+14 unique choices on planet Earth.

My computor has made more choices than that since I turned it on. I just broke the multiverse.

Without resolving quantum entanglement or anything else, string theory is unique in that it is the first widely publicized, marginally accepted theory that resolves no paradox and solves no problem.

Why the success?

The success of string theory demonstrates a general, even global frustration with physics. Once a physicist comprehends two or three paradoxes and a few of the implications of those paradoxes, the whole world of physics may seem to crumble. Some stand among the many books of their ivory tower and cry "These theories don't work. There are paradoxes everywhere. I need something else. I need something that makes sense!"

Strings, singularities, and dark matter will only go so far. To really get at these problems, you really have to dig down deep. You have to get to the core of the problems.

Supersonic Air Travel

Concord supersonic passenger jets (CSPJs) (1976 – 2003) came in with a boom.

Concord

Concord's supersonic fleet took off one year before the Russian supersonic fleet took off in 1977.

The Russian fleet was grounded in 1978 after less than one operational year and a crash at the Paris Air Show.

CSPJs, meanwhile, were promoted and hyped as the next generation of high-speed, low-cost air travel. The next greatest thing. The air frame that would make Boeing's 707 air frame the thing of yesteryear.

The plan must have looked great on paper.
- Fly at higher altitudes. At higher altitudes you get
  - Less wind resistance, so
    - Better fuel mileage
    - Higher-speed travel
  - You make less noise on the ground
  - You experience less gravity the higher you go, so

- Less weight
- Lower fuel costs
- Low cost, high-speed air travel!

It didn't work out that way.

Ticket prices ended up being notoriously high – 3,000 times the price of the cheapest tickets for the same routes.

What went wrong?

The Engineers' Paradox

CSPJs were not (initially) engineered for any increase in weight due to increase in altitude. Based on Cavendish's iron-core Earth model, the super-sonic fleet should experience less gravity the higher it gets.

Instead, Concord super-sonic jets ended up needing 20-30% more fuel than expected for a standard flight, and, contrary to engineering specifications, they ran more efficiently at <u>lower</u> altitudes.

Why?

Because at high altitudes, they weighed more than expected.

Airplanes are expected to weigh a *little* more at high altitudes, but that is not expected to happen due to gravity. Instead, when an airplane gains altitude, it is observed that the airplane will exist in lower air pressure. As the air pressure drops, the buoyancy of the airplane decreases, resulting in the additional experience of weight.

How much more? About 0.5% more - barely enough for an engineer to notice.

But Concord airplanes weighed as much as 5% more at high altitudes and fuel usage went through the roof. Engineers were hopelessly confused. Once again, gravity was all wrong. Their experience was paradoxically contrary to their model of a solid-core Cavendish Earth.

The increase in weight also stressed the physical components of the airplane. They were not designed to provide enough lift to compensate for the increased weight. Pilots were reported as refusing to operate at the altitudes designated for some flights.

Concord's people must have engineered their hearts out trying to resolve this paradox: Higher altitude was not more efficient to travel in. The original cost estimate for the Concord super-sonic program was £70 million. The program ended up costing over £1.3 billion.

Concord supersonic jets typically flew as low as possible and their sonic booms were a source of sore complaint wherever they roamed.

The overstresses of the improperly engineered aircraft became most apparent on July 25, 2000, when Air France Flight 4590 went down killing all one hundred and eleven people on board and another four people who were on the ground. That event grounded half of the supersonic fleet permanently.

Supersonic passenger jets were then only allowed to fly over the ocean and were banned in most of the countries that had not banned them already.

In 2003 the rest of the Concord supersonic fleet was retired. All six of them.

Sixteen years after the retirement of that fleet, the resolution to the engineers' paradox still remained unimaginable to most scientists: The Cavendish model of Earth is wrong.

The Boeing 707 air frame is still the international standard of large airplane design both militarily and commercially.

Airbus is Boeing's primary competitor. Modern airplanes by Airbus resemble Boeing's 707 air frame.

The Acceleration of Gravity

Airbus A318 Airbus.com 2019

The acceleration of gravity on the surface of Earth is 22 miles (35 kilometers) per hour per second. That is the speed of the downward acceleration experienced due to gravity. That is the net downward force of gravity. Running head-long into a wall at 22 miles per hour exerts the same amount of force as standing on the ground for one second.

If Newton's perpendicular model of gravity were accurate, then the speed of an object's horizontal travel would not change the angle of gravity. No matter how fast you go, gravity would still be pulling you straight down.

The experiments as Los Alamos establish a very different angle of gravity.

As shown in the illustration below, when an object is in motion horizontally, it defeats gravity to the rear and assists gravity in the forward direction. This results is an increase in weight proportional to speed up to about 360 kilometers per hour. The net total increase in weight is about 3.5%. The forces of gravity perpendicular to the direction of travel are not affected.

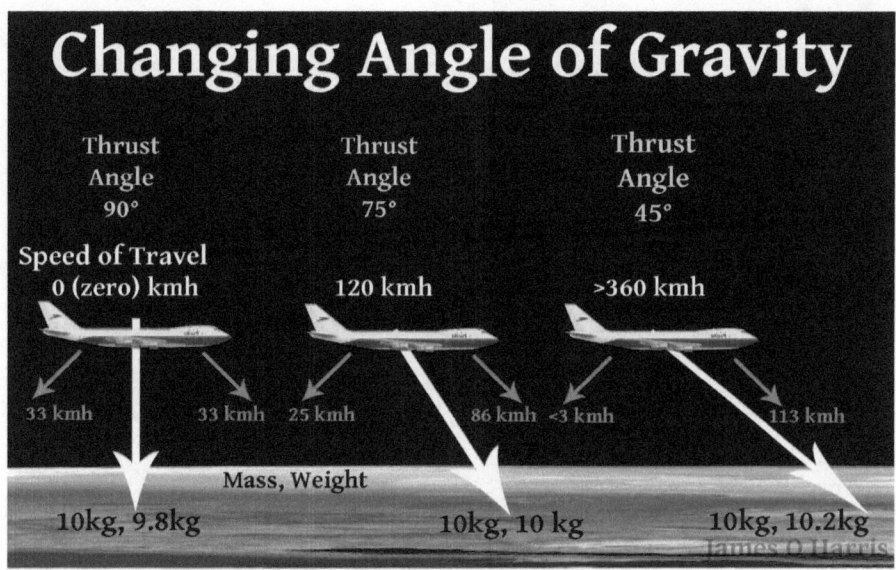

Changing Angle of Gravity

A small percentage of gravity is truly vertical gravity. Most gravity is non-vertical and averages out to 45°.

Gravity at such a low angle indicates a big difference between gravity and weight.

$$1 \text{ Earth g (little g for Earth gravity)} = \text{Mass} * \text{Newtons}$$
$$\text{Mass} = 101.9 \text{ kg};$$
$$1 \text{ Earth g} = 9.81 \text{ Newtons}$$
$$g = 9.81 \text{ N} * 101.9 \text{ kg}$$
$$\text{Weight} = 100 \text{ kg};$$

A 101.9 kilogram mass experiencing one Earth g weighs 100 kilograms on Earth's surface. How much gravity occurred?

To figure that out, we have to take the dominant angle of gravity – 45° for Earth – and use that angle's sine to calculate the net downward vertical force exerted by gravity.

$$\text{Weight } 100 \text{ kg} = \text{Gravity} / (1 - \text{sine of angle of gravity})$$
$$45° \text{ Sine} = 0.707$$
$$(1 - 0.707) = 0.293$$

$$100 \, kg = Gravity / 0.293$$
$$Weight \, of \, 100 \, kg = Gravity \, of \, 341 \, kg$$

On the surface of Earth, a mass of 101.9 kilograms weighs 100 kilograms and experiences 341 kilograms of gravitational force, shown below.

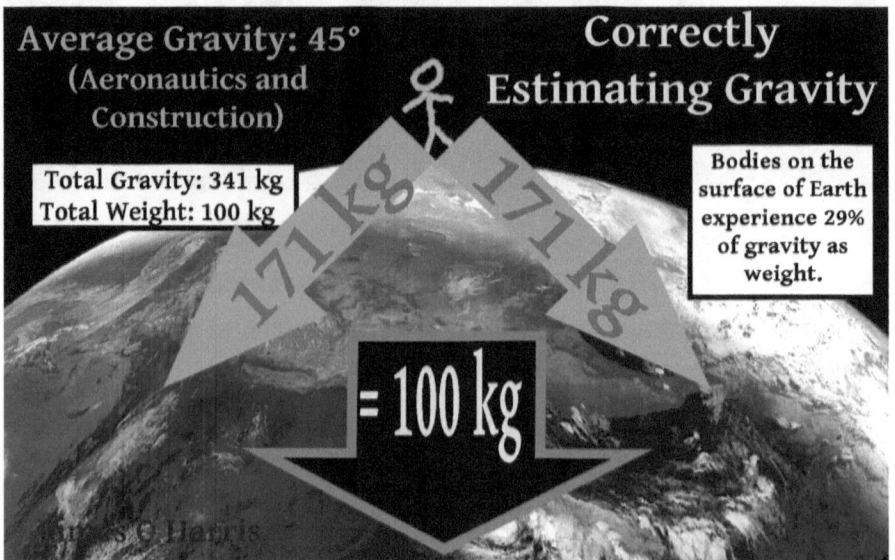

Now let's take that back to the Cavendish experiment. Cavendish calculated a 100 kilogram weight as experiencing 100 kilograms of gravity. If only 29.3% of gravity is demonstrated as weight, what does that do to the Cavendish results?

First off, that means that Cavendish was not calculating for the value of G. Cavendish was calculating something else.

Force = the downward aspect of gravity (29.3%

$$* \, gravity) * mass_1 * mass_2 / radius^2$$

That is not Newton's gravity equation.

Maybe Cavendish knew that. Maybe that is why Cavendish stressed that he had not established the value of G.

With the new information provided by the Los Alamos experiments, a new constant for gravity can be calculated, called the James Newton or ℕ.

$$6.67E\text{-}11 / 0.293 = ℕ\ 2.27E\text{-}10$$

Another component of the Engineer's Paradox is altitude. According to Cavendish Earth theory, the farther away you get from Earth's center, the less gravity you will experience.

Since Earth's mass is primarily near the surface, that is not true. As seen in this illustration, the weight of a 101.9 kilogram mass changes proportional to the angle of gravity. At zero degrees it weighs zero. At 45°, it weighs 29.3 kilograms. At 50° it weights 22% more. At 60° it weighs 71% more.

101.9 kg mass
= Weight
at various angles

90°
98%
100 kg

60°
50%

50°
36%

50.0 kg

60°

Average Angle of
Surface Gravity:
45°

35.8 kg
29.3 kg

45°
29%

50°

45°

0°
0.0 kg

0°
0%

James O Harris

So when Concord flew their airplanes higher and higher to escape gravity, the were going in the wrong direction. Instead of weighing less and less, they weighed more and more.

Now they just don't fly at all.

Paradox Null
~~Cavendish Earth~~
~~Engineer's Paradox~~

The insights made available through the Los Alamos aviation experiments should have utterly rocked the scientific world. Instead, they went by as if unnoticed.

Every year hundreds of scientists now travel in 707-type airplanes to go to conferences where they debate theories of a Cavendish Earth – an Earth very different than the one those airplanes fly over.

Our Impossible Galaxy

The Milky Way galaxy became impossible on February 13, 1974. That day, astronomers Robert Brown and Bruce Balick, from the National Radio Astronomy Observatory, discovered a black hole at the middle of our galaxy – a supermassive black hole – called Sagittarius A*. (Where an asterisk appears in a name like Sagittarius A*, it is pronounced as 'star'.)

Sagittarius A* is perhaps the largest and/or most massive object in the Milky Way galaxy. Sag A* is also right in the middle of the galaxy. The Milky Way effectively orbits Sagittarius A*.

What was special about Brown and Balick's observation was that this huge, obviously super-heavy mass was not white. It was not blue or red. It did not look like a star. It was a black hole, and if Sagittarius A* weighed nearly much as later calculated, then the nature of gravity had completely defied Newton and Einstein both. The Milky Way galaxy should be flying apart in every horizontal direction, and all galaxies should be crashing into one another.

In 1974, Frank Sherwood Rowland (1927 – 2012) and Mario José Molina-Pasquel Henríquez (1943 –) at the University of California at Irvine reported that chlorofluorocarbons and other long-lived organic compounds may have a destabilizing effect on the ozone layer. They demonstrated that CFCs chemically react with ozone, and that that destroys ozone, resulting in O_2, CO_2, and other chemicals instead.

The truth of the matter is that ozone chemically reacts with almost everything.

CFCs are also very light, so CFCs, the researchers reasoned, could float up into the upper atmosphere, destroy Earth's ozone layer, and end all life on Earth.

Fears would run high ten years later when the ozone hole was discovered.

The world of science now hinged on the assumptions of Copernicun gravity, flying photons, Cavendish Earth, gravity equality, 6.67, strange gases, random evolution, Darwin's timeline, neutral gravity, Helmholtz heat, constant light speed, Lorentz compression, flying electrons, nothing space, orbital electrons, Copernicun mass, floating continents, constant planet size, Eddington limitocity, omnipotent nothing, infinite photons, stellar fusion, Russian Expansion, doppelmatter, cosmic hatching, none of this exists, geocentric creation, salvation by ozone, white holes, dark matter, EPR, Higgs boson, microwave bang, and that CFCs can end all life on Earth.

Hawking Holes

In 1974, Stephen Hawking (1942 – 2018) presented his theory that black holes emit radiation. This concept later became known as *Hawking Radiation*.

Hawking radiation, in concept, is an attempt to overcome the *Information Paradox*. Classic black holes, which I will call *Hawking holes*, suck in matter, electricity, and light and nothing can ever escape. That obviously violates black body radiation laws and also the thermodynamics law of input=output.

Hawking also wrote the *Second Law of Black Hole Dynamics* which states that Hawking holes can never get smaller. The uncertainty principle of quantum mechanics, however, dictates that photons and electrons which blink in and out of existence would inevitably escape the black hole, rendering the Second Law of Hawking Holes false.

Robert Brown, Bruce Bilick, Slipher, and others had observed that x-ray radiation is emitted from the region of Sagittarius A*. In response to this, Hawking proposed that matter and photons are converted by Hawking holes into gamma-rays and x-rays and are then emitted away from the Hawking hole.

So, what were Hawking's gamma-rays and x-rays composed of?

Hawking's theory is that Hawking holes have such immense gravity that light – in the form of flying photons – cannot escape the gravitational influence of a Hawking hole.

In the Einsteiniverse, gamma radiation and x-ray radiation is *also* thought to be composed of photons.

Why would gamma photons radiate away from Hawking holes while photons of light cannot escape?

The world of science now hinged on the assumptions of Copernicun gravity, flying photons, Cavendish Earth, gravity equality, 6.67, strange gases, random evolution, Darwin's timeline, neutral gravity, Helmholtz heat, constant light speed, Lorentz compression, flying electrons, nothing space, orbital electrons, Copernicun mass, floating continents, constant planet size, Eddington limitocity, omnipotent nothing, infinite photons, stellar fusion, Russian expansion, doppelmatter, cosmic hatching, none of this exists, geocentric creation, salvation by ozone, white holes, dark matter, EPR, Higgs boson, microwave bang, CFC death, and that gamma-ray photons are governed by different physical laws than light-wave photons in the vicinity of Stephen Hawking's black holes.

In 1975, Paul Dirac published a 69-page work titled *General Theory of Relativity* which explained and expanded Einstein's relativities in English. This became the namesake of Einstein's 1915 theory set.

In 1979, Alan Harvey Guth (1947 -), a junior particle physicist at Cornell University, developed the concept of cosmic inflation.

1978
Nobel Prize
in Physics

Pyotr Leonidovich Kapitsa
"for his basic inventions and discoveries in the area of low-temperature physics"

Arno Allan Penzias and Robert Woodrow Wilson
"for their discovery of cosmic microwave background radiation."

Scientists had been very disappointed at that point in history at not finding the origin of a big bang. Hubble had predicted that when the catalog of galaxies grew large enough, that the source of the motion would become apparent. Instead, astronomers, obeying, Hubble's Law, found that it all started right here on planet Earth. At Cornell University in this case.

It had been 38 years since Hubble had disclaimed the idea of an expanding universe.

Yet Alan Guth proposed that the whole universe is inflating instead of exploding big bang style. Guth argued that – according to Hubble's Law – if you went to any distant galaxy and looked outwards, you would see expansion in every direction. He proposed that Hubble's Law does not make the Earth the center of the big bang. Instead, the whole universe is the center of the big bang.

Great. Right. So, um... where did it start?

In this model there is no starting point. This model – which was meant to support big bang theory – contradicts big bang theory. In big bang theory there was a cosmic egg or singularity, and it hatched from a single point. Contrary-wise, in cosmic expansion theory, every point is the starting point.

The world of science now hinged on the assumptions of Copernicun gravity, flying photons, Cavendish Earth, gravity equality, 6.67, strange gases, random evolution, Darwin's timeline, neutral gravity, Helmholtz heat, constant light

speed, Lorentz compression, flying electrons, nothing space, orbital electrons, Copernicun mass, floating continents, constant planet size, Eddington limitocity, omnipotent nothing, infinite photons, stellar fusion, Russian expansion, doppelmatter, cosmic hatching, none of this exists, geocentric creation, salvation by ozone, white holes, dark matter, EPR, Higgs boson, microwave bang, CFC death, Hawking holes, and that the big bang began at planet Earth _and_ everywhere else.

Expansion theory leaves many unanswered questions; "What is expanding?" Are the galaxies flying apart or is the underlying structure spreading? If the underlying structure is spreading what is the speed of light constant relative to? If the underlying structure is spreading, why aren't galaxies spreading internally?

In 1984, Paul Dirac published his last paper, titled *The Inadequacies of Quantum Field Theory,* where Dirac expressed his disgust over the infinities plaguing the field.

Dirac wrote, in part,

> *"How then do they manage with these incorrect equations? These equations lead to infinities when one tries to solve them; these infinities ought not to be there. They remove them artificially."*

Dirac went on to distance himself from Heisenberg uncertainty,

> *"Just because the results happen to be in agreement with observations does not prove that one's theory is correct. After all, the Bohr theory [of the hydrogen atom] was correct in simple cases. It gave very good answers, but still the Bohr theory had very wrong concepts."*

Paul Dirac died later that year at the age of 82.

Ozone Hole

In 1978 NASA had sent up Nimbus-7 – a satellite with an ozone detector. In 1984, Pawan Bhartia reported the first finding of the *ozone hole.*

Bhartia's findings were based on satellite surveys but the Nimbus detectors proved too weak to detect any possible *changes* to the ozone layer during that mission and gathered virtually no polar data.

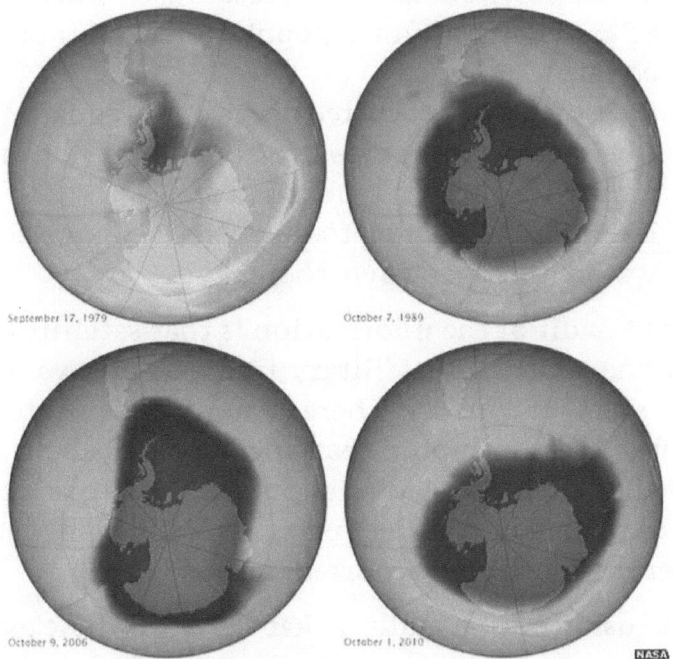

Independent of Bhartia, the British Antarctic Survey team including Farman, Gardiner, and Shanklin used ground-based antarctic equipment to measure the south pole's ozone layer. They reported finding the ozone hole in May of 1985 in *Nature* magazine.

No ozone hole exists anywhere else on Earth.

James O Harris

How the ozone hole occurred can be described in one of two distinctly different ways: It was suddenly found or it suddenly happened.

According to the popular account, the ozone did not exist in 1978, was faintly evidenced in 1981, was a full-blown hole by 1984, and has not appreciably changed since then. The ozone hole is interpreted as evidence of ozone depletion. That depletion is attributed to the concurrent increase in the production of chlorofluorocarbons.

These ideas came from the propositions of Sherwood and Molina. Those propositions fail to address major issues.
- 95% of Earth's human population exists north of the equator.
- 98% of CFC use occurred north of the equator.
- Air in northern temperate zones does not mix with air in the southern Antarctic zone.
- If CFCs had anything to do with the ozone hole, why would it happen down there?

A different audit of the information is that satellites did not fly over the south pole. Military interests always influence where a satellite will fly. There is no military intelligence concerning the south pole that anyone cares about. Efforts to view the north pole led missions on trajectories over the south pole, though, and the first effective view of the ozone hole over the south pole occurred in 1984.

The conclusion of that audit is that the ozone hole existed in 1984 and there is no reason to believe that it did not always exist. It looks just like it always has. It changes with Earth's distance from the Sun. It changes with the spin of the Earth. It changes with improved viewing technology.

This report out of the Netherlands indicates that the production of CFCs took a downfall around 1990. Although CFCs have nothing to do with the ozone hole – and although ozone is considered a pollutant – the notion that CFCs destroy the ozone layer was canonized and resulted in an international treaty known as the *Montreal Protocol.* That agreement largely banned the production and use of CFCs in North America and Europe. The Montreal Protocol was passed in 1987 and phased out most lightweight CFCs before the end of 1996.

Twenty-three years later the ozone hole remained visually unchanged.

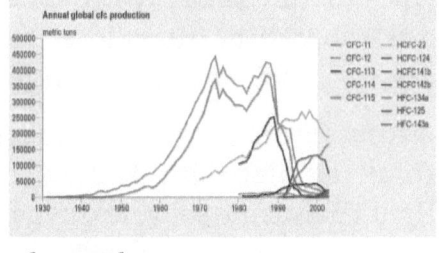

Popularly held beliefs concerning the ozone layer overlook the most fundamental implications of the findings of Sydney Chapman.

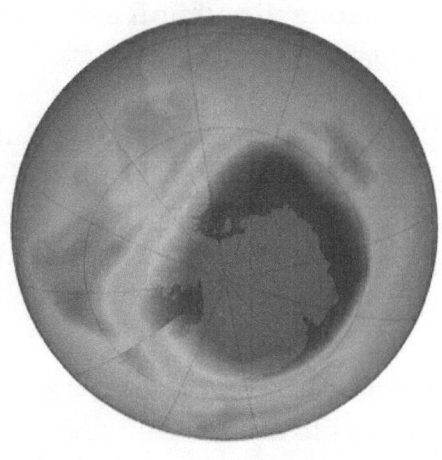

Chapman found that ultra-violet light *creates* ozone from *atmospheric oxygen.* Among other things, this means that the more UV light there is, the more frequently ozone will occur.

This well-known fact is employed in virtually every modern hot tub, where electrical discharges are used to convert atmospheric oxygen into ozone, and that ozonated air is then injected into the water to kill microbial life forms. The device is called an *ozone generator.*

So if you stripped away all of Earth's ozone tonight, new ozone would begin to occur as soon as the Sun came up tomorrow.

But someone might point out that there isn't much oxygen up that high – 15 to 35 kilometers in altitude.

Ozone exists that high up because it floats. Ozone – being far more energetic than oxygen (O_2) – takes up a lot more space.

Ozone weighs merely 1/6th as much as atmospheric oxygen. So when ozone occurs, it tends to go upwards.

The farther up in the atmosphere it travels, the more ultraviolet light it receives, so the ozone is kept as ozone and ascends to the ozone layer.

Once in the ozone layer, ozone continues doing what it always does, absorbing ultraviolet radiation and re-emitting that energy primarily as infrared light.

Ozone forms because atmospheric oxygen is highly reactive to ultraviolet light. If the ozone doesn't stop the ultraviolet light then atmospheric oxygen will.

Another neglected matter in the ozone discussion exists as the *polar jets* of Earth. Just like black holes, the Sun, and other planets, Earth ejects more than 70% of the gases it will lose – 181,440 kilograms of the 259,200 kilograms vented daily – through jets which erupt from the north pole and south pole.

The solar wind consists of primarily hydrogen and helium. The solar wind is literally an ordinary wind.

Earth emits the likeness of a solar wind consisting of primarily hydrogen and helium. Earth's 'little solar wind' is literally an ordinary wind.

Earth emits 95 million tons of gases as 'little solar wind' every year.

About 70% of Earth's 'little solar wind' emerges at Earth's polar jets which extend to 50,000 kilometers before they being deflected by the solar wind of the Sun

These outbound gases originate below the surface and those winds blow a lot of oxygen and nitrogen into Earth's upper atmosphere. Then, while hydrogen and helium continue outbound, oxygen, carbon dioxide, water, and nitrogen mostly fall back towards the Earth. Those winds induce static electricity. Electrical discharges cause ozone and the aurorae to form.

So the polar jets are ozone factories.

The polar jets propel gases from the surface of the Earth to an altitude of around 50,000 kilometers or 49,950 kilometers above the ozone layer.

But what's with the south pole? There is *never* an ozone layer there.

The south pole has never had an ozone layer because the south pole is the positive (+) magnetic pole of Earth.

Biases within Earth's magnetic field affect the stability – the very existence – of ozone.

The key to the ozone hole is that ozone does not form in an electromagnetically positive-bias environment. Ozone is a negative-bias electromagnetic structure.

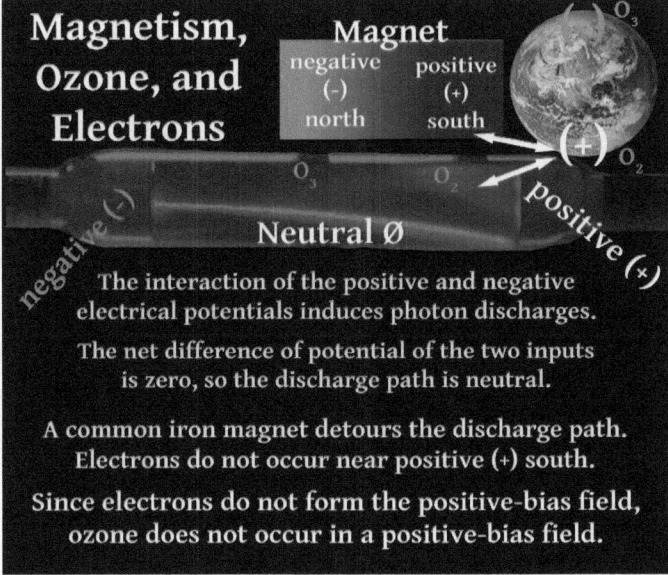

Magnetism, Ozone, and Electrons

	Magnet	
negative (-)	positive (+)	
north	south	

O_3

O_2

positive (+)

negative (-)

Neutral Ø

The interaction of the positive and negative electrical potentials induces photon discharges.

The net difference of potential of the two inputs is zero, so the discharge path is neutral.

A common iron magnet detours the discharge path. Electrons do not occur near positive (+) south.

Since electrons do not form the positive-bias field, ozone does not occur in a positive-bias field.

In the illustration above, **electrons** do not form in the cathode tube near the positive (+) pole of the magnet above. Electrons and ozone occur in the same place – away from the positive pole of the magnet, away from Antarctica.

The north pole of the magnet has little apparent influence. On Earth, the ozone layer is slightly thicker than average near the north pole.

When discussing the ozone hole and ice sheets, it is important to recognize that information concerning the polar regions is very scarce. Both the arctic and antarctic circles are *no-fly zones* enforced by the United States military.

Is the Earth's magnetic field strong enough to affect the production of ozone? Look at your compass. Just kidding. You have GPS. But if you have an old-fashioned disc compass, your compass has a north indicator which is positive-bias. Your compass's north pole indicator is actually the south pole of your magnet so that it points to Earth's magnetic north pole.

The magnetic north pole of Earth is the negative-bias (-) pole of the Earth. Your compass's north pole indicator (+) is attracted to that.

In the southern hemisphere, compasses barely work. The north (-) field of the Earth (and every magnet) is far stronger than the south (+) field.

The north field is so much more potent than the south pole that early magnet theory only had one pole, the other pole being neutral.

Some say that this belief was overcome by Michael Faraday who, being a blacksmith, magnetized a horse-shoe and demonstrated that both ends picked up iron shavings. If only one pole was active, he reasoned, that would not happen. Whether this actually happened or not, the logical argument is sound. This same experiment also proves that the north (-) pole is more powerful than the south pole (+). The north pole of a given magnet is typically capable of lifting 25% more weight than the south pole is.

Because the north pole is so much more potent than the south pole, compasses are ineffective in the southern hemisphere.

In the southern hemisphere, compasses do not point north. They do not point south, either. The compass's magnetic attraction to the Earth's north pole remains stronger than the opposite end's attraction to the south pole. That attraction is not useful for guidance, however, because when a compass is in the southern hemisphere, the northern hemisphere is not physically north of the compass. The north pole is below the compass instead – way below the horizon.

In the southern hemisphere, north is more than 45º degrees downwards, so you are facing north in mostly every direction.

That is one way of expressing how that Earth's magnetic field is quite powerful. It also extends about 100,000 kilometers into space. The Earth's magnetic field is more than strong enough to prevent the formation of ozone at the south pole.

The world of science now hinged on the assumptions of Copernicun gravity, flying photons, Cavendish Earth, gravity equality, 6.67, strange gases, random evolution, Darwin's timeline, neutral gravity, Helmholtz heat, constant light speed, Lorentz compression, flying electrons, nothing space, orbital electrons, Copernicun mass, floating continents, constant planet size, Eddington limitocity, omnipotent nothing, infinite photons, stellar fusion, Russian expansion, doppelmatter, cosmic hatching, none of this exists, geocentric creation, salvation by ozone, white holes, dark matter, EPR, Higgs boson, microwave bang, CFC death, Hawking holes, omnibang, and that the ozone is diminishing at the south pole because of human activity near the north pole.

In 1986, Bohr published some of his final thoughts on physics in *American Scientist Vol. 74*. Of all matters of physics, Bohr was most eagerly troubled by time.

> *"Time, among all concepts in the world of physics, puts up the greatest resistance to being dethroned from ideal continuum to the world of the discrete, of information, of bits. ... Of all obstacles to a thoroughly penetrating account of existence, none looms up more dismayingly than 'time.' Explain time? Not without explaining existence. Explain existence? Not without explaining time. To uncover the deep and hidden connection between time*

and existence ... is a task for the future."

Welcome to the future. Welcome to *The Age of Paradox Null.* Niels Bohr was right. Unlocking the keys to time unlocks the keys to existence.

Cosmic Microwave Map

In 1992, the COBE (Cosmic Background Explorer) took detailed measurements of the *Cosmic Microwave Background Radiation (CMBR or CMB)* and found anisotropy – differences in intensity based on direction.

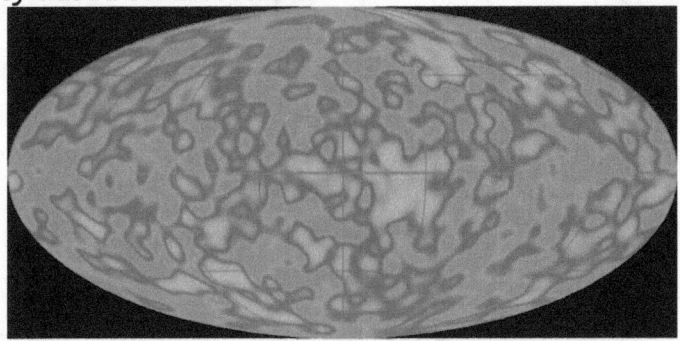

The most significant challenge when it comes to photographing the CMB is getting the Milky Way galaxy out of the way. The Milky Way galaxy and <u>all other galaxies</u> must be be excluded from the CMB since the COBE was looking for *background radiation* only.

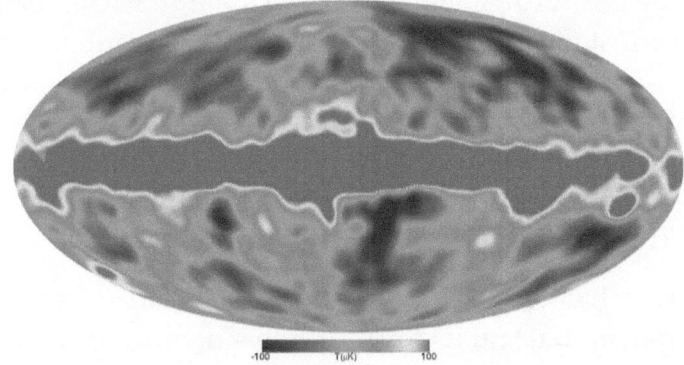

Above is the COBE image by NASA without removing the Milky Way galaxy.

In 1995, the first anti-hydrogen was produced by a team led by Walter Oelert (1942 –) at CERN. Oelert is a Professor at the

James O Harris

Juelich Research Center in Germany.

This discovery spawned new faith in big bang theory, theoretically allowing for the existence of the opposite of matter.

Nobel Prize in Physics

Russell A. Hulse and Joseph H. Taylor Jr.

"for the discovery of a new type of pulsar, a discovery that has opened up new possibilities for the study of gravitation."

That conceptually allows that every-thing could occur from nothing as long as opposite of that stuff also occurred. The 'nothing' from which big bang matter emerged was a singular-ity. The 'something' that emerged was matter.

The first half of the equation is 'something' – stuff, matter, atoms, and so on. We know that exists because, well, Descartes said it best, "I think, therefore I am."

Something =

The second half of the equation is 'nothing'. Nothing is the sin-gularity which Lemaître demonstrated that Einstein's e=mc^2 predicts. This is followed up with the idea that the universe is expanding and, therefore, everything occurred from nothing.

Something = Nothing

Something is obviously wrong with that equation, so another variable must be introduced to offset this imbalance. The third part of the equation is 'anti-something'. Anti-something would offset something.

Something + Anti-something = Nothing
(Something + Anti-something = 0) = Nothing
0 = Nothing

This allows that zero equals nothing and would place the uni-verse at mathematical harmony with singularity theory.

And that mathematical harmony would be a major proof of Einsteinian relativity as well. If that theory accurately pre-dicted antimatter, then that would bring relativity back into harmony with thermodynamic theory where input equals output.

Input = output

The imbalance which Lemaître capitalized on for his cosmic egg theory is the same imperfection which violates the thermodynamic law. Einstein had addressed this imperfection with a *cosmological constant* to *avoid* an exploding or imploding universe. Einsteins had also attempted to connect his theory of *white holes* to his cosmological constant to no avail.

The Hubble Telescope: New Out-sights

By 1995, the big bang theory had evolved from happening 200 million years ago involving extra-galactic nebulae (Lemaître) to happening 2 billion years ago involving 32 galaxies (Hubble) to 4 billion years ago (Darwin). While that happened, the galaxies multiplied, and the Earth proved itself 4.5 billion years old.

In December of 1995, the Hubble Telescope took the the *Hubble Deep Field* (HDF) image of galaxies 8 billion light-years away.

This NASA-produced photograph was subtitled, "Hubble Deep Field Image Unveils Myriad Galaxies Back to the Beginning of Time."

At that shocking point in history big bang birthday was pushed back from 4 billion years to 8 billion years. It was then authoritatively declared that nothing beyond the 8 billion light-year horizon could exist. Period. Because it would have to have traveled faster than the speed of light.

But the HDF included galaxies with red-shifts exceeding 6 – traveling away from Earth (according to Hubble's Law) at six times the speed of light. This notion did not become popular until after the Hubble Ultra Deep Field (HUDF) image was released a few years later.

The background noise in the Hubble Deep Field is *Cosmic Microwave Background Radiation,* which, as seen later, is more galaxies.

Dark Energy

Dark energy theory emerged in the 1930s in light of Fritz Zwicky's dunkle materie, but retained little interest in the astronomical community until 1998. It then had a renaissance in 2011 and passed away in November of 2016.

In 1998 the Supernova Cosmology Project and the High-Z Supernova Search Team published independent reports which stated that type Ia supernova, as observed in galaxies far, far away, were getting redder.

According to both of those reports, that meant that the speed of the expansion of the universe was accelerating.

Reports of 'accelerated expansion of the universe' sparked

conversations about *dark energy*, which, based on those reports, was occurring at an increasing rate and driving the universe apart.

Those findings were met with a lot of skepticism in the scientific community. It was incompatible with classic big bang theory that expansion would be accelerating. It should be slowing down instead.

And based on that idea, the universe is dying.

The universe is dying in that paradigm because the ever-increasing totality of dark energy results in ever-increasing acceleration of galaxies far, far away from each other.

Also in the Einsteiniverse, stars destroy matter, convert that matter into energy, and broadcast that energy.

So in this expanding Einsteiniverse, energy is continually produced and broadcast out into empty expanding space. The energy, therefore, becomes dilated and diluted and the temperature of the universe continually falls.

And since that energy originates as matter, the quantity of matter in the universe is decreasing.

This adds up to the science-fictional conclusion that the universe will expand into a freezing empty nothing – that the universe will become a de Sitter universe.

In 1998 Michael Turner (1949 -), with a PhD in Physics from Stanford University, coined the term *Dark Energy*. The term 'dark energy' was a take-off of Zwicky's dark matter in that it was invisible and broke all laws of physics. However, dark energy was very different from dark matter in that dark energy did not have attractive gravity, but repulsive gravity or 'anti-gravitational effects'.

The world of science now hinged on the assumptions of Copernicun gravity, flying photons, Cavendish Earth, gravity equality, 6.67, strange gases, random evolution,

Darwin's timeline, neutral gravity, Helmholtz heat, constant light speed, Lorentz compression, flying electrons, nothing space, orbital electrons, Copernicun mass, floating continents, constant planet size, Eddington limitocity, omnipotent nothing, infinite photons, stellar fusion, Russian expansion, doppelmatter, cosmic hatching, none of this exists, geocentric creation, salvation by ozone, white holes, dark matter, EPR, Higgs boson, microwave bang, CFC death, Hawking holes, omnibang, crazy ozone, and that a spontaneously occurring anti-gravitational invisible virtual substance is exploding more than before.

Dark energy was created to resolve the paradox left behind by Baade when he reduced Hubble's constant speed. That paradox, simply stated, is that if the galaxies are moving so slowly, that motion would be overcome by gravity and the galaxies should all be crashing together.

But they're not.

2000 – 2019

Wilkinson's MAP

Following up on the identification of anisotropy by the COBE mission, in 2001 the Wilkinson Microwave Anisotropy Probe (WMAP) was launched on a 6-month mission to more precisely map the CMB. Photographic replication of the findings became an international hit.

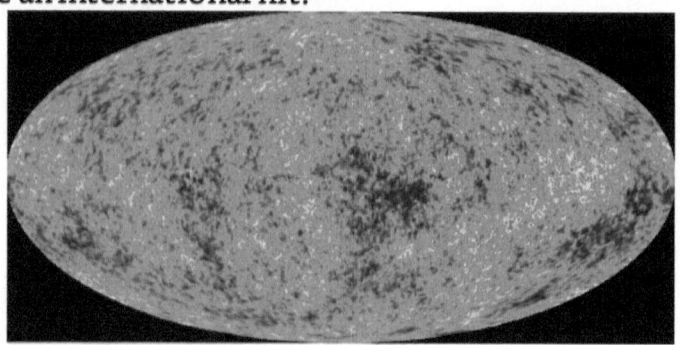

Like COBE, this NASA-produced image took rigorous work to generate because all known galaxies, including the Milky Way and its stars – had to be removed to generate this extra-galactic view of the CMB.

WMAP looks very different if the Milky Way and other galaxies are not removed.

Without removing known galaxy objects, *Example 19* shows that the Milky Way Galaxy is delivering far more microwave radiation to the Earth than the rest of the universe is.

This NASA-produced poster shows the results of each of the twelve different spectra that the WMAP probe measured. The Milky Way is by far the most powerful source of radiation across all tested frequencies.

If galaxies emit microwave radiation so powerfully, why does anyone think that CMB is not originating from galaxies?

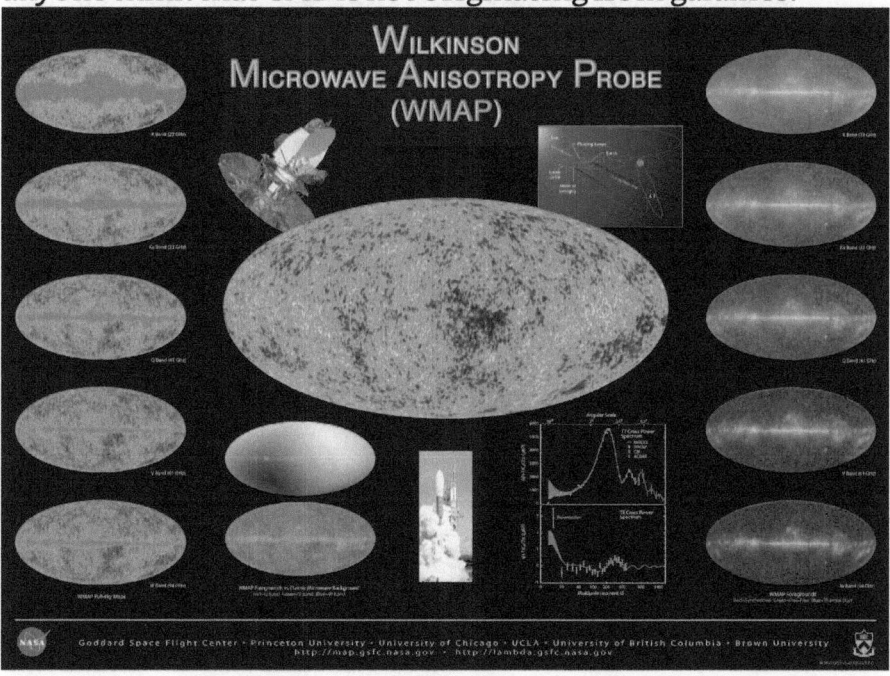

The world of science now hinged on the assumptions of Copernicun gravity, flying photons, Cavendish Earth, gravity equality, 6.67, strange gases, random evolution, Darwin's timeline, neutral gravity, Helmholtz heat, constant light speed, Lorentz compression, flying electrons, nothing space, orbital electrons, Copernicun mass, floating continents, constant planet size, Eddington limitocity, omnipotent nothing, infinite photons, stellar fusion, Russian expansion, doppelmatter, cosmic hatching, none of this exists, geocentric creation, salvation by ozone, white holes, dark matter, EPR, Higgs boson, microwave bang, CFC death, Hawking holes, omnibang, crazy ozone, dark energy, and that CMB

does not come from galaxies.

Jack Hills

Meanwhile, down under, in 2001, Simon A. Wilde, John W. Valley, William H. Peck, and Colin M. Graham reported in *Nature* that the Jack Hills region of Australia contained strata which had been there for 4.56 billion years.

In this study, the researchers used uranium dating. Uranium dating is done by finding uranium in ancient typically volcanic rock strata and examining the rock it is embedded in for trails of lead which have been left behind as uranium deteriorated. The proportion of uranium to lead indicates how long the uranium has been embedded in the stone up to 4.65 billion years.

These studies focus on the uranium isotope 235U. Uranium isotope 235U has a half-life of 8 billion years. As 235U breaks down, a certain proportion of it becomes lead.

The problem is that there isn't very much 235U. Only 0.71% of Earth's uranium is 235U. Another problem is guessing at how much of a given uranium sample was *initially* 235U. The imperfect guesswork as to a batch of uranium's original 235U content makes radio-dating back beyond 4.65 billion years too unreliable to be considered scientific.

This model of calculation has other imperfections as well. This model relies on the assumption that all 235U was formed at the same moment is cosmological history. If that is not true, and if 235U is continually generated, than many assumptions concerning radio-dating must be reexamined.

For the next 13 years, the age of the strata from this Jack Hills report was commonly reported as 3.65 billion years old. That age was reported because that is within the margin of error on some of the measurements, and because the prevailing notion at the time was that Earth was only 4 billion years old.

Simon et al. found that to be an interesting point of conversation.

"This indicates that at 4.4 Gyr ago there were already intermediate to granitic, high-d18O **continental rocks** to contaminate the magma in which grainW74/2-36 grew. **The existence of liquid water at 4.4 Gyr ago** could have important implications for the evolution of life. Microfossils as old as 3.5 Gyr are known. Metasediments and carbonaceous materials with biogenic carbon isotope ratios as low asd13Cà228Ω are known at 3.8 Gyr ago. Zircon crystal W74/2-36 is over 500 Myr older than this organic matter." (emphasis added)

Simon et al are making, in that last sentence quoted, 'The zircon crystal was 500 million years older,' a reference to a softly spoken scientific fact: All of the carbon-related compounds used to radio-date Earth's most ancient strata *are associated with biological activity.*

Originally, radio-dating using carbon isotopes was heralded as the next great advance in evolution theory. Finding isotopes which only occur due to biological activity, scientists speculated, would reveal when life on Earth began. Carbon isotopes cease to be isotopes at very high temperatures, so no carbon isotopes could exist before Earths fiery birth if everything was molten metal. The isotopes they were working with would not be present in strata which existed before life.

As radio-dating technology advanced, however, major incongruity between evolution theory and carbon-dating evidence arose. One example is a carbon isotope which has a half-life, it was determined, of 17.5 billion years. Worse than that, not all of that isotope is still radio-active. That indicates that life has existed on Earth and in Earth's strata for at least 1 trillion years.

As with uranium 235U, this isotope is only a small proportion of the carbon available, so radio-dating was only effective, initially, back to about 8 billion years ago. From this discovery, the age of the universe/big bang was moved back to 8 billion years. That adjustment happened in the field of geology about 40 years before astronomy moved big bang theory back that far. Modern technology enables radio-carbon dating to reach back 22.5 billion years.

Throwing Away the Evidence

Since the radio-carbon isotopes (associated with biological activity) have a definite half-life, the age of the isotopes is determined by how much deterioration has occurred – through determining what proportion of the isotope is *not* radioactive any more.
-    If they find a high percentage, it's a new crystal.
-    If they find a low percentage, it's an old crystal.
-    If they find a very low percentage, it's a very old crystal – up to 4.45 billion years old.
-    If they find no radiation, the crystal may be more than 4.5 billion years old, and they throw it away.

The key to radio-dating is working with radioactive material. If there isn't any, the inferred evidence is commonly overlooked. If you find a sample with what would be radioactive chemicals, and those chemicals are not radioactive any more, perhaps the age of the stone is greater than those methods can measure.

The key to this science is that the isotopes were encapsulated within a zircon crystal. That prevents contamination. No matter what happens or happened to that crystal since the radio-active carbon isotopes were locked up, they remain isolated from any contamination, intrusion, or loss of chemicals (besides hydrogen and helium).

The process of locking up radio-active isotopes inside of zircon crystals only happens in volcanoes where the right combination of temperature, pressure, and biomatter exist.

But if a crystal containing carbon isotopes is older than 4.45 billion years, it is difficult to figure out how old that is, so we throw it away.

CREATIONISM

Astronomers and religionists have always wanted less time for creation. Some push for a time table as short as 6,000 years and the group as a whole has stood firm against exceeding 13.7 billion years. Evolutionists at large lost interest in the creation debate when Earth's age got to 4 billion years. Geologists typically want more time for creation.

Gravity Paradoxes

The reason why astronomers have always wanted less time is that the further back big bang theory birthday is pushed, the slower the Hubble constant has to be. The slower the Hubble constant is, the bigger the gravity paradoxes become and the more dark matter or dark energy you need. That in mind, for astronomers, the ideal Hubble constant is around 150 kilometers per second per mega-parsec. At that speed, the need for dark energy is minimal and the maximum age of the universe is around 6 billion years.

Evolution Paradoxes

On the evolutionist side, Darwin himself in *Origin of Species* speculated that the mutationary evolution process would take many millions of times longer than Lemaitre's big bang theory allowed – that translates to *trillions* of years.

There are at least three problems with evolution theory. First origin paradoxes, natural selection, and fossil extinction.

The Paradox of First Origin

In the Beginning, there was mud.

The first problem with evolution theory is *The Paradox of First Origin.* The problem there is that the simplest life forms have flagellum to get around. A flagellum is biological electrical motor with a wireless remote control which adjusts the shape of the *filament* (propeller) to control the direction of movement.

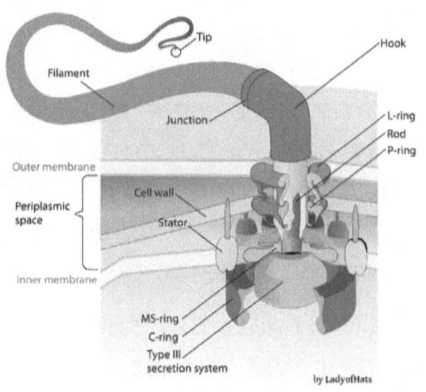

Seriously, you can't get one these by electrifying some mud. And if that looks complicated, try zapping one these into existence: A fully autonomous, self-replicating complex protein: DNA.

Of course, DNA won't succeed without a life support system – the cellular structure including a breathing mechanism, feeding apparatus, that propeller, and 200+ different chemicals composed of 50+ unique elements.

All of that has to form in the same place at the same time as a single, cohesive unit.

What Evolution? The Greenhouse Approach.

Stromatolites now grow as stromatolites grew 3.5 billion years ago. Megalodons grew 23 million years ago as great white sharks grow now. Crocodiles lived 200 million years ago as they do now. Volcanoes erupted at least 5 billion years ago as they do now.

The most notable change is that all of these things have gotten smaller over the last 65 million years.

What if the atmosphere has grown thinner?

Animals, fish, and insects raised at four times Earth's normal atmospheric pressure grow, in the case of fish, to three times their normal length. Plants raised in 1.5 atmospheres of pressure require 200% more carbon dioxide and higher temperatures. Their photosynthetic rate skyrockets.

The phenomenon controlling which life forms exist on Earth is natural selection.

Dinosaurs cannot exist as large creatures in a thin atmosphere. At lower pressures megalodons grow only as large as great white sharks. In a thin atmosphere, land animals, birds, and small-winged dinosaurs weigh twice as much as they do in a super-critical fluid atmosphere of the same gases.

And volcanoes?

Higher surface pressure means that a volcano requires more internal pressure to erupt. If a volcano requires more pressure to erupt, then more energy will build up before that eruption happens, and that makes for bigger eruptions.

The polar ice caps?

Those did not exist until – at the utmost extreme and based on plankton-related Antarctic deposits as evidence – until 33.6 million years ago. That is the indirect evidence used by the Andalusian Institute of Earth Sciences (IACT) in a 2013 report.

Direct evidence provides shorter timelines. On December 17, 2017 Ohio State University reported in "Oldest ice core ever drilled outside the polar regions,"

> "The oldest ice core drilled in the Northern Hemisphere was found in Greenland in 2004 by the North Greenland Ice Core Project and was dated to roughly 120,000 years, while the oldest continuous ice core record recovered on Earth to date is from Antarctica, and extends back 800,000."

Until less than one million years ago, there was no persistent Antarctic ice sheet. If the atmosphere was significantly denser back then, Earth's atmosphere was also much deeper. A deeper atmosphere results in a greenhouse effect whereby more clouds reflect more heat. That raises temperatures and also levels temperatures across the planet somewhat. In this case, that also kept the poles much warmer.

Case in point: In 2011 Certa IA et al reported in Naturwissenschaften that Antarctica supported huge, herbivorous sauropods 65 million years ago.

Extinction of Fossils

Speaking of eruptions, extinctions, and ice sheets, the whole surface of the Earth was once a molten wasteland. <u>But not all at once.</u>

The whole surface of the Earth has experienced volcanic activity. Hard to believe? The Siberian Traps erupted numerous times and over the course of two of those eruptive cycles covered an area as large as the continental United States.

Volcanoes are typified by basins – sides of mountains blown out during an eruption – and springs of water which emerge near mountain peaks. Ski resorts such as Breckenridge and Vail in Colorado are arranged in volcanic basins. Across Colorado's Front Range, springs of water emerge near virtually every mountain-top. The whole American mountain range – from Peru to Alaska – formed volcanically. Mountains consist of sequential volcanic deposits, resulting in layering which does not form level surfaces, nor ever did.

For at least five billion years, volcanoes have been erupting, sinkholes have been sinking, and rain has been falling. That destroys most fossil records, buries some fossils beyond recovery, and forms some fossils – especially in the mares (volcanic mud-flows) that often result.

The bigger an animal is, the less likely it is that that animal will be buried and fossilized. If a fossil is formed, the larger that fossil is, the more rapidly it will be broken up by settling, erosion, or biological activity. The bigger the fossil, the faster that fossil goes extinct.

That greatly biases the fossil record and may misguide evolution theory.

Radioactive Paradoxes

Geologists can demand far more time for creation than evolutionists do. One of those reasons is tellurium (Te). Tellurium was first discovered in 1782 in a gold mine in Kleinschlatten, Transylvania and has two radioactive isotopes: 128Te, which has a half-life of 2.2E+24 years and 130Te which has a half-life of 8.2E+20 years.

On Earth, 31.74% of tellurium is 128Te, 34.08% is 130Te.

Stable isotopes of tellurium include 126Te – 18.84%, 125Te –

7.07%, 124Te – 4.74, and the lightest isotope 120Te. 34.18% of tellurium is stable.

These isotopes are interesting, in part, because tellurium's atomic number is 52. In the classic helium atomic model, tellurium should be stable at 104Te – an isotope which does not exist.

If all that tellurium formed at the same time and in a radioactive state, and half of the stable tellurium – 15.9% – was 128Te, then two thirds of one half life has occurred (68%) indicating that the tellurium on Earth is about 1.5E+24 years old.

That is 1,500,000,000,000,000,000,000,000 years old (1.5 septillion). That is almost 1.5 septillion years longer than 13,700,000,000 years (13.7 billion).

One of the logical processes of determining when 'creation' occurred was with the use of uranium. The long and short of that is that scientists determined that 235U has a half-life of about 7 billion years and determined that more than half of it has broken down – the exact same method outlined for tellurium above. From that, the age of the Earth was established as 8 billion years.

Based partly on that, big bang theory was pushed back to 8 billion years.

But that method is flawed. That method assumes that all radioactive substances occurred at the same time.

The world of science now hinged on the assumptions of Copernicun gravity, flying photons, Cavendish Earth, gravity equality, 6.67, strange gases, random evolution, Darwin's timeline, neutral gravity, Helmholtz heat, constant light speed, Lorentz compression, flying electrons, nothing space, orbital electrons, Copernicun mass, floating continents, constant planet size, Eddington limitocity, omnipotent nothing, infinite photons, stellar fusion, Russian expansion,

doppelmatter, cosmic hatching, none of this exists, geo-centric creation, salvation by ozone, white holes, dark matter, EPR, Higgs boson, microwave bang, CFC death, Hawking holes, omnibang, crazy ozone, dark energy, magical CMB, and that radioactive substances on Earth originated all at the same time.

That assumption – whether they all occurred at a big bang, generally, or they all occurred at a supernova locally – is not borne out by evidence.

On Earth, 99.274% of uranium is 238U with a half-life of 4.5 billion years. All uranium is radioactive. That makes it fun to theorize over. Since it is all radioactive, and since that, in Einstein's model, that means that it is atomically evaporating, that means that uranium just goes away or changes to lead.

So uranium-based theories can be based on lead deposits or other abstract information. Uranium theories can justify Earth as a hotter planet or even a molten planet based on how much uranium is added to a theory.

The tellurium model does not allow for that guesswork.

But both of those models are flawed.

When and where do radioactive elements occur?

If all the radioactive substances occurred shortly after a big bang and the universe is 1.5 septillion years old, uranium would not exist. 238U would have gone through over 122 quadrillion half-lives by now. It would all be gone.

If uranium occurred in a supernova, it would blow up the way it does in an atomic bomb. Uranium cannot come from supernovae.

Then look at radium. Radium also has no stable isotopes. Like uranium, all radium is radioactive. Radium isotopes have half-lives of 3.6 days to 1,600 years.

If Earth is 8 billion years old, the longest-lasting radium has gone through 5 million half-lives. There would not be any radium on this planet. And 3.6 days? If radium does not occur locally on planet Earth, how could we possibly have <u>non-synthetic</u> isotopes here which have half-lives of just 3.6 days?

What are half-lives anyhow?

A half-life is the period of time during which the high-frequency emissions of an unstable atomic substance reduce to one half.

This occurs because either

- Half of the atomic structures have become rearranged such that they no longer emit high-frequency radiation, or
- Half of the atomic structures have dissipated as by fission.

But how can someone possibly observe a half-life which occurs over a span of 2 million billion billion years?

The method of calculating non-observable half-lives is calculated, instead, by inference.

Using Faraday's folly in the form of $e=mc^2$, the observed energetic output is relativized to mass. In the case of tellurium, the energetic output per year is one half (half-life) of $1/2.2E$ $+24$ of the relativized mass of tellurium. Theoretically, the tellurium is fissioning very, very, slowly.

Radioactive output is due to atomic energy transformation (AET). More on that later.

Half-lives are due to nuclear rearrangement of atomic substances.

Radioactive substances have very high levels of energetic output. This relates to the amount of ambient potential (zero point energy) which the atom transforms into electromagnetic energy.

Everything always transforms energy. That experience is known as 'time'. But radioactive substances do *a lot* of that transformation.

What's wrong with that? That's like putting giant propellers on an ordinary windmill.. Sure, it'll produce more power, but only until it breaks or falls over.

Since we are talking in terms of proto-electromagnetic transformation, radioactive atoms may be thought of as having extended antennae. Antennae capture and transform ambient energy.

What happens when antennae generate *too much* power? They'll blow up, blow apart, or melt down. On an atomic level, that's like changing a tennis ball into a racket ball of similar size. The 'fuzz' of the tennis ball were the atomic 'antennae' responsible for the high levels of atomic energy transformation.

If the antennae melted down, the mass of the atom remained the same. If the antennae exploded away, the mass diminished slightly.

What else goes on in there?

The structures of radioactive substances vary wildly from uranium which is a very high-density nucleus to cesium which is a very low-density nucleus with wide-open spaces inside.

Due to the unusual structures of radioactive substances, and the high levels of AET, radioactive substances may cause nucleogenesis, forming other elements including hydrogen, helium, and lead.

Lighter elements may also form from heavier elements through *partial fission*.

Radioactivity is also infectious.

Exposure to high-energy radiation can alter atomic structures, forming those 'antennae' which then cause the affected atom to become radioactive – emitting high levels of energy at high frequencies.

When a human body is exposed to radioactivity, certain isotopes in the body become radioactive with half-lives ranging form several minutes to 17.5 billion years.

And that can be good for you.

The fact is that the human body manufactures and/or depends on many radioactive substances, including potassium, and every cell produces helium – a product of either fission or fusion.

In both thermonuclear generators and the human body, the most important chemical for the management of radioactivity is *tetrasodium borate*. Boron is a very low-energy atom which is used to control radiation.

Molten Paradoxes

A common creation myth is that Earth formed at extremely high temperatures. Several paradoxes surrounding that concept are discussed later, with the most striking paradox being that Earth would have blown away in the solar wind.

Illustration Boiling Points of Planet Earth demonstrates what Earth would be like at 5,000 Kelvin using the Cavendish model of an iron-core heavy planet Earth. At 5,000 Kelvin, about 0.5% of Earth's mass would be solid and liquid, the rest would be gaseous.

Earth's strongest gravity would occur at about halfway between the outer layer and the core and would total about 0.1 gs of gravity.

The Boiling Points of Planet Earth

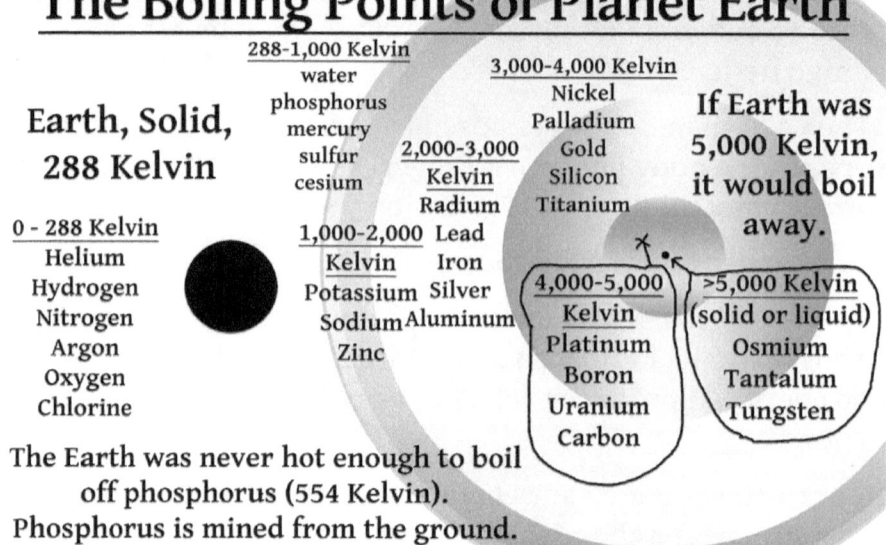

Earth, Solid, 288 Kelvin

288-1,000 Kelvin
water
phosphorus
mercury
sulfur
cesium

3,000-4,000 Kelvin
Nickel
Palladium
Gold
Silicon
Titanium

2,000-3,000 Kelvin
Radium
Lead
Iron
Silver
Aluminum

If Earth was 5,000 Kelvin, it would boil away.

0 - 288 Kelvin
Helium
Hydrogen
Nitrogen
Argon
Oxygen
Chlorine

1,000-2,000 Kelvin
Potassium
Sodium
Zinc

4,000-5,000 Kelvin
Platinum
Boron
Uranium
Carbon

>5,000 Kelvin
(solid or liquid)
Osmium
Tantalum
Tungsten

The Earth was never hot enough to boil off phosphorus (554 Kelvin).
Phosphorus is mined from the ground.

The center of Earth would be primarily osmium, tantalum, and tungsten. Everything else would form a huge, gradient atmosphere, making Earth about the size of Uranus.

Without a centralized mass, there would not be enough gravity to hold anything together. The Earth would not have formed from this ball of gases. The majority of the planet would have blown away.

And while we're looking at this model, consider what would happen as this ball cooled if it did stay together. When it cooled to 4,000 Kelvin, platinum, boron, uranium, and carbon would become the second, outer core. At 3,000 Kelvin, nickel, palladium, gold, silicon, and titanium condense as the third layer of the core.

None of these elements are ferrous (magnetic), so Earth's core cannot be magnetic in this model. Also in this model, all of Earth's platinum, gold, tungsten, nickel, and silicon are locked up in Earth's core.

That does not sound like an accurate description of Earth.

REWRITING THE STORY OF EARTH IN 5 MINUTES

These are the basics of the theories which the author has developed concerning Earth and its structure based on the angle of gravity, the spreading of Pangaea into continents, the scale of time during which that spreading occurred, the decrease of gravity, the decrease in the boiling point of water, the decrease of the weight of the kilogram, the extinctions of the dinosaurs, and other factors.

The Law of Expansion

The Earth's radius is increasing. Here is the new formula to calculate Earth's expansion.

$$\text{Radius new} = \text{radius old} + (\text{former expansion} * \text{expansion constant } \mathbb{C})$$
$$R_n = R_o + (ex_a * \mathbb{C})$$
$$\mathbb{C} \text{ Earth} = 1.15$$

The Law of Expansion applies to all celestial objects which are spherical or nearly spherical and/or all objects greater than 100 kilometers in radius. The expansion constant for each object is unique. Expansion applies most particularly to objects in binary orbit. Expansion occurs very rapidly within binary orbits.

Earth exists in a binary orbit with Luna. Here is a graph of the history of Earth's expansion.

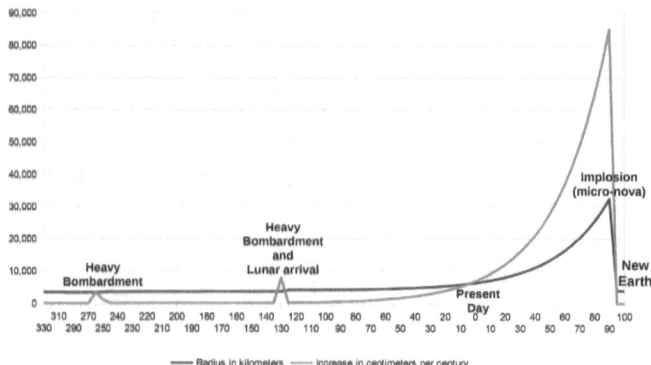

The expansion of Earth is due to the tidal effects of the Sun and, primarily, the Moon. The Moon arrived about 125 million years ago.

As Earth's center of gravity moves outward, Luna's orbit reflects that change, also moving outward.

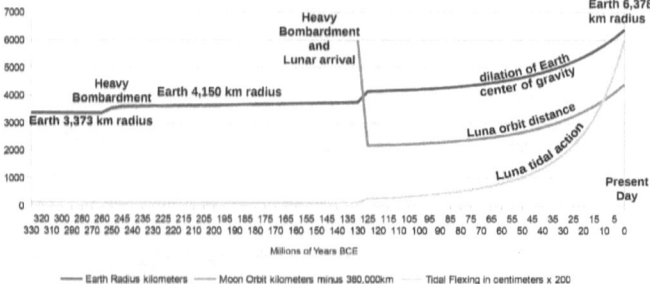

This trend ends in catastrophe. As Earth expands, the solid structure of Earth thins, resulting in implosion which will occur in about 90 million years.

As shown in the illustration below, Earth's expansion and the tidal effect of the Moon are exponentially increasing. The Earth will eventually collapse. At that point in time, Earth will have expanded to over 32,000 kilometers in radius and will be larger than Uranus.

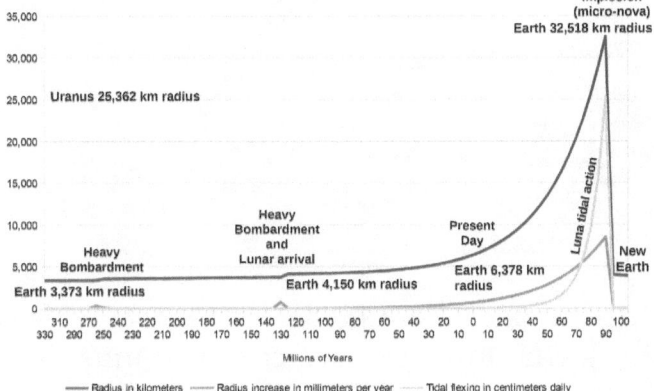

Then Earth will fall together and form a planet like the dinosaurs lived on.

As Earth's radius increases, gravity falls. When Newton designed gravity in 1687 there were 10 Newtons of gravity on Earth's surface. 333 years later there were 9.8 Newtons of gravity on Earth's surface.

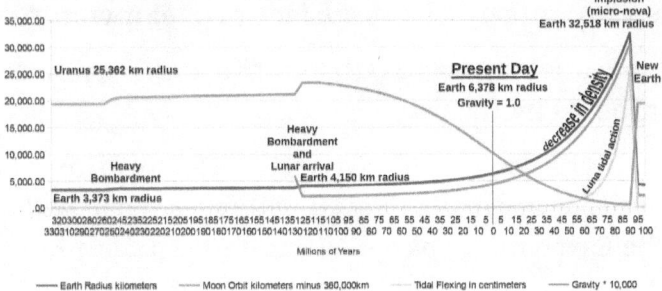

As gravity falls Earth's atmospheric pressure exponentially decreases. That has resulted in dynamic long-term natural selection which has extincted many creatures – especially the larger ones. The loss of atmosphere also allowed Antarctica to freeze.

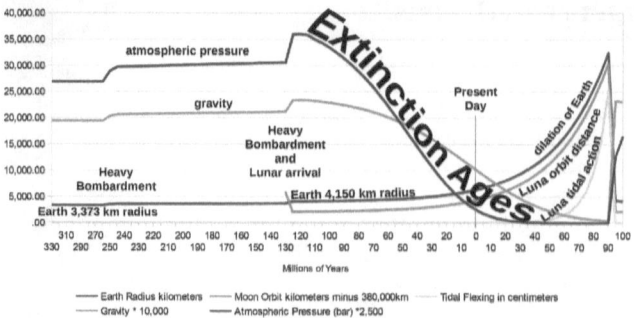

The extinction ages are a global and irreversible phenomenon. The extinction ages will continue unless something knocks the Moon out of orbit.

Local extinctions occurred relative to mega-eruptions and the resultant ice ages. The geological record frequently records similar creatures through a series of extinctions. These creatures typically re-inhabited the affected region between ice ages, typically migrating north. The southern hemisphere does not experience ice ages because ash sinks and snow does not pile up on the surface of ocean-water.

In 2019, oceans covered 80.9% of the southern hemisphere. That number is increasing.

But what if the Moon did *not* arrive 125 million years ago? Or what will happen after Earth collapses? Has the Moon destroyed the Earth in the past? Will it happen again?

If the Moon sticks around when Earth collapses, the projected cycle – the lifetime of Earth – is 225 million years.

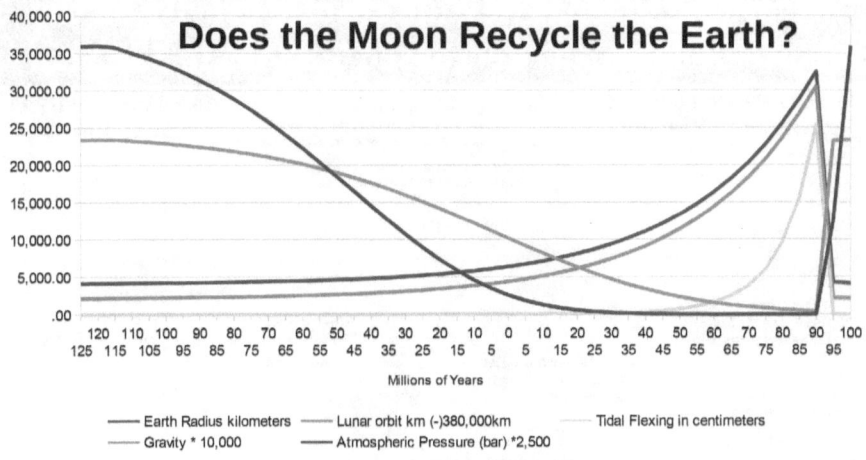

The Moon is expected to collapse and reform in about 25 million years and at intervals of about 80 million years.

Earth's atmospheric pressure is increasing. No. It's not. Earth's atmospheric pressure *appears* to be *increasing* because mercury loses weight as gravity decreases. This results in barometers developing smaller vacuums and, therefore, indicating *higher* pressure as gravity falls. As explained in the graphic below, mercury loses a lot more weight than water does as gravity decreases.

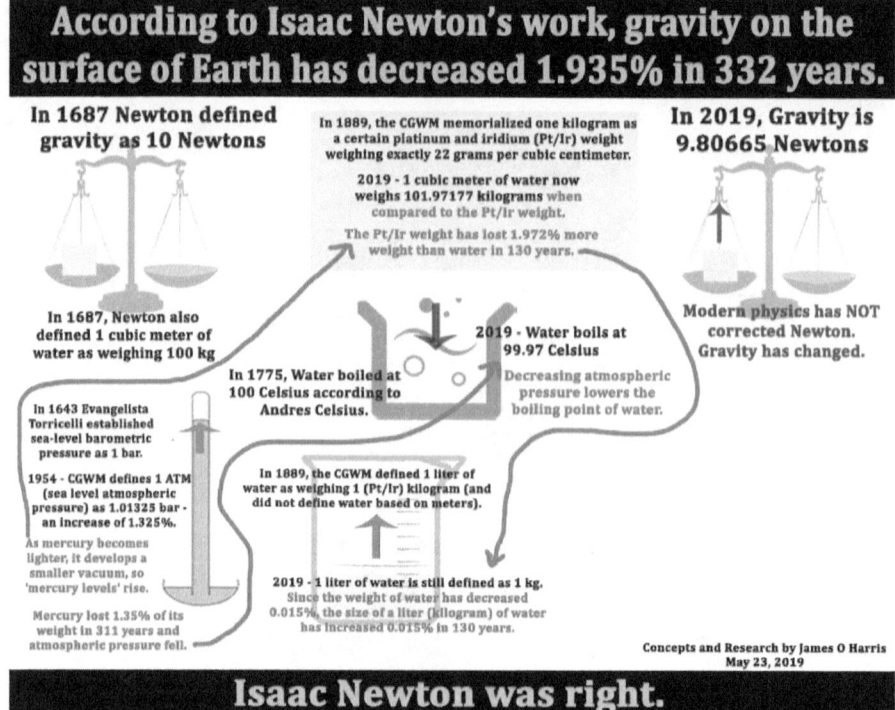

According to Isaac Newton's work, gravity on the surface of Earth has decreased 1.935% in 332 years.

In 1687 Newton defined gravity as 10 Newtons

In 1889, the CGWM memorialized one kilogram as a certain platinum and iridium (Pt/Ir) weight weighing exactly 22 grams per cubic centimeter.

2019 - 1 cubic meter of water now weighs 101.97177 kilograms when compared to the Pt/Ir weight.

The Pt/Ir weight has lost 1.972% more weight than water in 130 years.

In 2019, Gravity is 9.80665 Newtons

In 1687, Newton also defined 1 cubic meter of water as weighing 100 kg

2019 - Water boils at 99.97 Celsius

Modern physics has NOT corrected Newton. Gravity has changed.

In 1775, Water boiled at 100 Celsius according to Andres Celsius.

Decreasing atmospheric pressure lowers the boiling point of water.

In 1643 Evangelista Torricelli established sea-level barometric pressure as 1 bar.

1954 - CGWM defines 1 ATM (sea level atmospheric pressure) as 1.01325 bar - an increase of 1.325%.

In 1889, the CGWM defined 1 liter of water as weighing 1 (Pt/Ir) kilogram (and did not define water based on meters).

As mercury becomes lighter, it develops a smaller vacuum, so 'mercury levels' rise.

Mercury lost 1.35% of its weight in 311 years and atmospheric pressure fell.

2019 - 1 liter of water is still defined as 1 kg. Since the weight of water has decreased 0.015%, the size of a liter (kilogram) of water has increased 0.015% in 130 years.

Concepts and Research by James O Harris
May 23, 2019

Isaac Newton was right.

Supermassive Hawking Hole

On October 16, 2002, an international team of scientists led by Rainer Schödel of the Max Planck Institute for Extraterrestrial Physics released the results of a ten-year study of Sagittarius A*.

2002
Nobel Prize in Physics

Raymond Davis Jr. and Masatoshi Koshiba

"for pioneering contributions to astrophysics, in particular for the detection of cosmic neutrinos"

Riccardo Giacconi

"for pioneering contributions to astrophysics, which have led to the discovery of cosmic X-ray sources."

The study focused on an orbital known as S2*. The orbit of S2* is highly elliptical – like the orbit of a comet . When S2* went through perihelion (its shortest orbit distance), Schödel calculated its speed at 5,000 kilometers per second, indicating the mass of Sag A* as 3.7 million solar masses.

This was heralded as a major achievement for general relativity. For the mass of 3.7 million solar masses to exist within a sphere so small was seen as confirming Einstein's concept of singularity-based black holes – *Hawking Holes.*

For that much mass to exist in the area observed indicated that Sagittarius A* had an average density (not reported) of about $30g/cm^3$ – 1.55 times the density of gold – far less than the 'infinite' density predicted.

Although this discovery seemed to justify Einstein's theories, it also happened to create major problems in the world of astronomy. The behavior of the spins of spiral galaxies demonstrate that they do <u>not</u> orbit their centers gravitationally.

This problem brought *dark matter* into the galaxies.

Instead of dark matter existing as Zwicky's extra-galactic force of attraction, dark matter became an *inner-galactic* force of attraction instead.

Dark matter was employed to explain why spiral galaxies do not orbit their centers.

The world of science now hinged on the assumptions of Copernicun gravity, flying photons, Cavendish Earth, gravity equality, 6.67, strange gases, random evolution, Darwin's timeline, neutral gravity, Helmholtz heat, constant light speed, Lorentz compression, flying electrons, nothing space, orbital electrons, Copernicun mass, floating continents, constant planet size, Eddington limitocity, omnipotent nothing, infinite photons, stellar fusion, Russian expansion, doppelmatter, cosmic hatching, none of this exists, geocentric creation, salvation by ozone, white holes, dark matter, EPR, Higgs boson, microwave bang, CFC death, Hawking holes, omnibang, crazy ozone, dark energy, magical CMB, radio-birth, and dark matter version 2.0.

In 2002, Riccardo Giacconi won the Nobel Prize for work related to detecting x-rays coming from Sagittarius A* – from the heart of the Milky Way Galaxy.

In, 2004 the Hubble Telescope was focused to 13.7 billion light years away for the Hubble Ultra Deep Field image. The Hubble Telescope has a range of 500 billion light years.

Hubble Ultra Deep Field (HDF) by NASA

That is when bending the rules for the speed of light became popular.

All of these galaxies violated not only big bang law, but Einstein Law as well. If the big bang happened, for these galaxies to get this far away they would have had to have traveled faster than Einstein's light-speed speed limit.

2006 Nobel Prize in Physics

John C. Mather and George F. Smoot

"for their discovery of the blackbody form and anisotropy of the cosmic microwave background radiation."

So the big bang theory was changed such that the big bang exploded faster than the speed of light initially. The was possible in de Sitter space. There was no matter in de Sitter space, so that could travel just as fast as it wanted.

The new absolute maximum size/age of the universe was then

set to 13.7 billion light years.

The world of science now hinged on the assumptions of Copernicun gravity, flying photons, Cavendish Earth, gravity equality, 6.67, strange gases, random evolution, Darwin's timeline, neutral gravity, Helmholtz heat, constant light speed, Lorentz compression, flying electrons, nothing space, orbital electrons, Copernicun mass, floating continents, constant planet size, Eddington limitocity, omnipotent nothing, infinite photons, stellar fusion, Russian expansion, doppelmatter, cosmic hatching, none of this exists, geocentric creation, salvation by ozone, white holes, dark matter, EPR, Higgs boson, microwave bang, CFC death, Hawking holes, omnibang, crazy ozone, dark energy, magical CMB, radio-birth, more dark matter, and that violating the laws of physics is okay when you've violated them already.

In 2009, Zwicky and Baade's theory of supernovae as the origin of cosmic rays was addressed in a report given at the International Cosmic Ray Conference (ICRC). At that conference, Hague, J. D delivered *Correlation of the Highest Energy Cosmic Rays with Nearby Extragalactic Objects in Pierre Auger Observatory Data*. Hague determined that there was no correlation between the incidence of cosmic rays and supernovae.

On July 9, 2005 two MAGIC (Major Atmospheric Gamma-ray Imaging Cherenkov) telescopes (2004 –) detected that gamma-ray photons arriving from blazar Markarian 501 were not all arriving at the same time.

Photons at different energy levels arrived at different times, suggesting that some photons traveled more slowly than others and contradicted the notion that the speed of light was constant. Photons with energies between 1.2 and 10 TeV arrived 4 minutes after those in a band between 0.25 and 0.6 TeV. The average delay was 30 ±12 ms per GeV of energy of the photon.

$$c_2 = c - (\text{photon energy} / 2E+17 \text{ GeV})$$

This delay, researchers speculated, occurred due to certain energies of photons interacting differently with something which impeded their travel. The impediment, they suggested, was Wheeler's quantum foam.

Minor signal delays are consistent with how prisms split light and how rainbows occur. The MAGIC telescopes are ground-based devices in the Canary Islands. The signals they receive have traveled through Earth's atmosphere where, among the gases, wave-particle transformation delay has occurred, slowing some wave-lengths more than others.

But delayed by 4 whole minutes?

Subsequent testing and experimentation have not supported that light is significantly delayed as suggested by the MAGIC observations or the proposed equation. MAGIC probably received a comparable wave from a comparable event which occurred four minutes after the first event. For example, a meteor fell and four minutes later another meteor fell, with the first one burning up at a lower temperature than the second one.

The concept of a quantum foam which delays or scatters light is not generally accepted by the scientific community.

In 2009, Gillessen et al. reported in *The Astrophysical Journal* a sixteen-year study of the supermassive black hole at the middle of our galaxy – Sagittarius A*. Like Schödel, Gillessen also focused on observations of S2*, and did so with better equipment. Gillessen calculated the speed of S2*'s as reaching 7,650 kilometers per second.

Since the apparent speed was so much higher, the apparent mass of Sag A* was also higher. Gillessen et al. estimated the mass of Sag A* at 4.31 million solar masses.

This number is strange for a number of reasons. First of all, the identified size of Sagittarius A* is 393,152 solar volumes, but weighing 4.31 million solar masses SagA* was being estimated as 11 times denser than the Sun.

The Sun weighs 1.37 grams per cubic centimeter, so Gillessen's finding may be extrapolated to indicate that the average density of Sag A* is 15 grams per cubic centimeter. That's about 70% the weight of platinum and 90% heavier than iron. Not much of a black hole.

Whether Sagittarius A* is 3.7 million or 4.31 million solar masses is immaterial. Either one of those numbers ruin our whole galaxy. Those numbers ruin the entire universe.

Galaxy Gravity Problems

It had been assumed since the days of Isaac Newton (1642-1727) that our galaxy orbited a central point like planets orbiting a star. That has never been backed up very well. Local stars and star clusters do not travel differently than each other with respect to Sagittarius A*.

If the Milky Way galaxy orbited Sag A*, then the objects closest to Sag A* would orbit Sag A* very quickly, and objects far away from Sag A* would orbit very slowly.

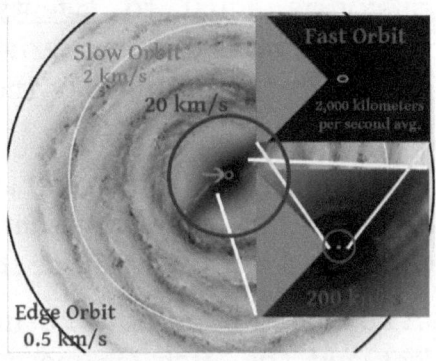

If our galaxy orbited Sagittarius A* directly and gravitationally, then the orbits of the Milky Way galaxy would look something like what is shown in *Illustration 96.*

But things do not work out that way. There are, instead, 3 distinct spin zones. 1: The Core – The innermost 1% of the Milky Way galaxy including Sag A*. 2: The Bar – 14% of the galaxy exists as a bar rotating at one cohesive rate. 3: The Disc – 80% of the galaxy is in the Disc, which rotates at one cohesive speed like a spinning Frisbee.

The other 5% of a galaxy's mass is outside of the disc and beyond the central galaxy.

Instead, what really happens is that the majority of a galaxy spins like a Frisbee, the outer parts travel much faster than the inner parts do.

That demonstrates that Sagittarius A* is not what's holding our galaxy together. Far from it, Sagittarius A* only controls a very small region.

But since galaxies are assumed to orbit their centers, the force described as holding the outer galaxy together was described as dark matter.

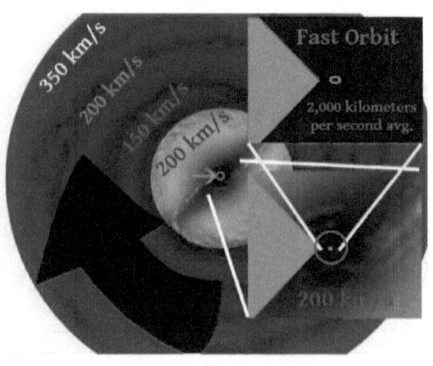

The calculated mass of the dark matter in the Milky Way galaxy is equal to <u>twenty times</u> the mass of all material objects in the Milky Way including black holes, stars, and planets.

In 2010, the Antihydrogen Laser Physics Apparatus (ALPHA) team at CERN generated the first known anti-hydrogen and conducted tests to establish the properties of anti-matter.

In 2011, The Supernova Cosmology Project and the High-Z Supernova Search Team collectively won a Nobel Prize for their 1998 reports in identifying via supernovae that the expansion of the universe was accelerating. The awards went to Saul Perlmutter, Adam Riess, and Brian

2011
Nobel Prize in Physics

Saul Perlmutter;
Brian P. Schmidt and
Adam G. Riess

"for the discovery of the accelerating expansion of the Universe through observations of distant supernovae."

P Schmidt.

Those are the same three scientists who designed supernova theory in the first place.

The presumably independent natures of the Supernova Cosmology Project and the High-Z Supernova Search Team was suspect. The major members of the two teams had divided their efforts only a few years prior to the famous reports.

One might wonder what 'Benefit to Humanity' this purported discovery entailed.

In the field of quantum mechanics, meanwhile, there was a light somewhere in a tunnel.

I Dream of Higgs

Experiments in 2012 at the Large Hadron Collider (LHC) and at the Conseil Européen pour la Recherche Nucléaire (CERN) concluded – for the first time in over 30 years of experiments – the 'detection of phenomena which *may* indicate the mechanisms suggested within the theory of the Higgs boson.'

This huge non-discovery was heralded as proving QM's Higgs boson theory correct. Did it work? No.

If the mechanisms of gravity were revealed through the supposed near-discovery of a Higgs boson, then those discoveries would have discovered, defined, and resolved dark matter and dark energy. No such advancements occurred.

In the seven years following the 'discovery' of the Higgs boson, further experiments did not support the headlines and further investigations met failure.

So there they were, in a world full of paradoxes, and 325 years after Isaac Newton they knew nothing more about gravity than he did.

2012 also boasted a new purported proof of Einstein.

Gravity Waves

The famous signal pictured below was captured by LIGO – The Laser Interferometry Gravity wave Observatory. LIGO facilities are huge interferometers each several kilometers long on each leg.

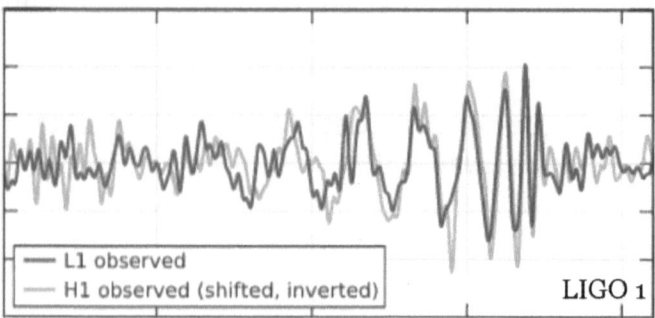

LIGO detects gravity waves the same way that Michelson-Morley tried to detect aether drag – by detecting uneven light speed across two different directions.

The way that LIGO detects gravitational waves is depicted in the graphic below where the red waves are gravity wave peaks and gray waves are gravity wave troughs. Red arrows are light-wave positive (+) peaks, black arrows are light wave negative (-) peaks.

The red waves accelerate light. The gray waves decelerate light.

On the horizontal axes, this has a net-zero effect on how long transmission takes. On the vertical axes, these changes significantly affect the rate of the transmission of light. These irregularities result in light *not* arriving at opposition at the output meter, resulting in positive or negative output.

(Distortion of light across the horizontal axes not visually depicted.)

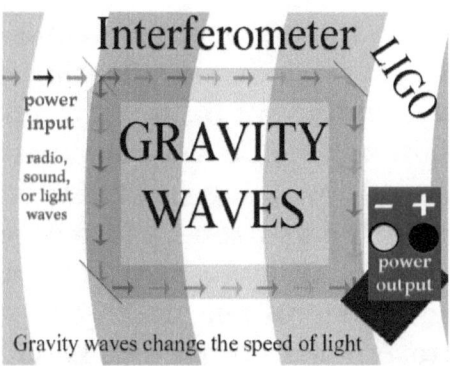

Gravity waves change the speed of light

Since the LIGO is built with Einsteinian physics in mind – where the speed of light is constant – when LIGO sees results, it interprets the irregularity not as a changing speed of light. Instead, the signal distortion is interpreted as changes to the physical size of the tunnel containing LIGO.

Strangely, although the Einsteinian gravity waves change the physical dimensions of LIGO's tunnel, those waves do not change the physical dimensions of the light waves. Those are exempt for some reason.

And since the light-waves are exempt from gravity wave dilation, the difference in the relative size of the light waves and size of the tunnels results in irregular output.

In 2012, there were two LIGO facilities several hundred miles apart – L1 and H1. For a gravity wave to be accepted as a gravity wave, both facilities had to find get the same signal at the same time. LIGO had failed at getting gravity waves for over 50 years.

In 2012, they finally got one.

The gravity wave picture was broadcast around the world and the Nobel Prize in Physics was forthcoming.

The world of science now hinged on the assumptions of Copernicun gravity, flying photons, Cavendish Earth, gravity equality, 6.67, strange gases, random evolution, Darwin's timeline, neutral gravity, Helmholtz heat, constant light

speed, Lorentz compression, flying electrons, nothing space, orbital electrons, Copernicun mass, floating continents, constant planet size, Eddington limitocity, omnipotent nothing, infinite photons, stellar fusion, Russian Expansion, doppelmatter, cosmic hatching, none of this exists, geocentric creation, salvation by ozone, white holes, dark matter, EPR, Higgs boson, microwave bang, CFC death, Hawking holes, omnibang, crazy ozone, dark energy, magical CMB, radio-birth, more dark matter, physics is optional, and universal warpage by gravity wave.

The gravity wave observed is commonly described as indicating the following chain of events 0.3 seconds:

This signal was described as 'waves of the warping of space-time due to the binary orbit of two very massive objects.' This was heralded as a long-sought-after proof of the correctness of Einstein's theory of general relativity.

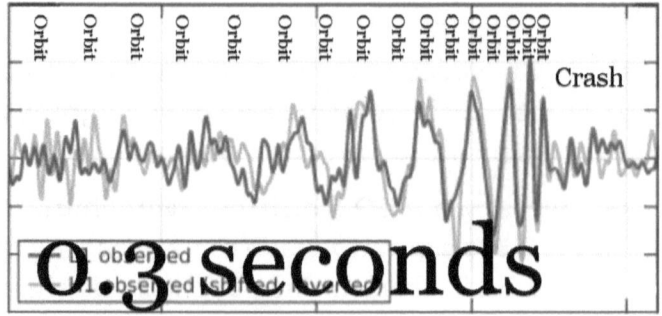

This signal – this rather well-defined, mutually-received signal – was purportedly the result of two black holes in a closing binary orbit spinning into a mutual collision which marked the end of the signal.

There is something wrong here. A closing binary orbit would generate a sinusoidal signal. A nice, round sinusoidal signal.

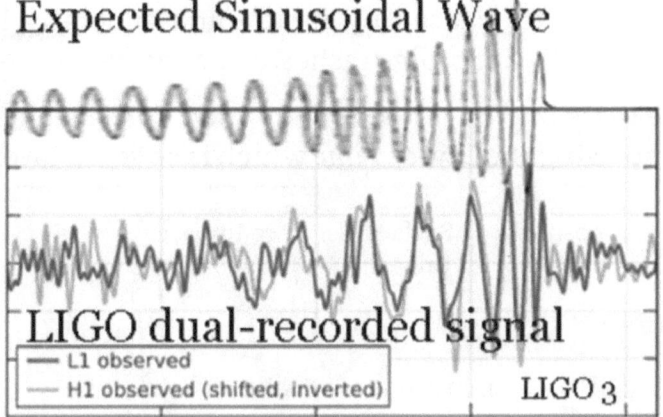

Expected Sinusoidal Wave

LIGO dual-recorded signal

— L1 observed
— H1 observed (shifted, inverted)

LIGO 3

The gravity wave signal timing from LIGO is all wrong. The peaks are in the wrong places; whole cycles are missing.

Researchers claim that those irregularities are due to signal loss and distortion.

Really?

- Look at the *clarity* of the LIGO chart! How can you have that many common peaks, and have whole signal cycles *completely missing?*

- The signal does not go to zero, as if something interfered with the signal halfway through.

- If something *was* interfering, how can something block *half* of a signal, leaving the bias high, and then, low, sequentially and contrary to the true signal?

- And how can all of that happen in 0.3 seconds?

The author's interpretation of this signal is characteristically different.

Gravity waves are caused by fission and generally indicate inter-stellar collisions, big crashes, and supernovae.

The time scale in *Illustration Gravity Wave Conclusion* has been calculated using the *Law of Time* which is covered later.

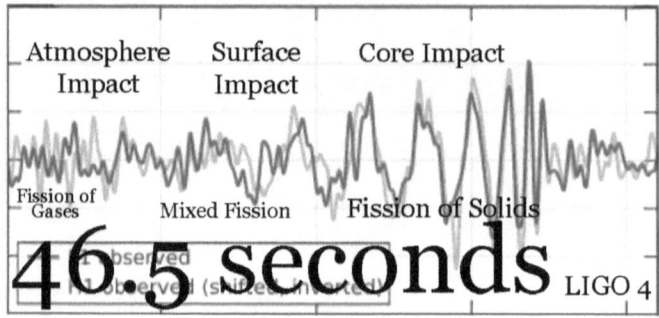

A fundamentally inconceivable error in the LIGO model is "How can to objects orbit each other with such tremendous velocity?" There are apparently 17 orbits occurring in 0.3 seconds in that model.

The objects involved were 40 and 20 solar masses respectively. If those objects were in a binary orbit at the speeds indicated then they would have flown away from each other and never collided.

The theory that these 'gravity waves' occurred from an orbit is absurd.

Another interesting point is that the theory of gravity waves violates the Einsteinian theory of a universally constant speed of light. If physical distances change, then the speed of light must also change. Otherwise, the speed of light is not constant. Oh yeah, and LIGO would not detect gravity waves.

The year 2012 also delivered a crippling blow to Einstein's physics.

XDF: A Deeper Look

The Hubble Extreme Deep Field (XDF) was released on September 25, 2012.

The XDF is an image of galaxies ~48 billion 'light years' away. A specially interesting aspect of the XDF is that it was taken in ultraviolet light. The galaxies photographed do not demonstrate any obedience to 'Hubble's Law' of red-shift.

Case closed on big bang theory? Hardly.

Sticking to their foregone conclusion, scientists kept the 13.7 billion light year rule and have now changed the rules for both light speed *and* space-time. Now space-time it thought to be expanding independent of flying photons.

That way, like a boat paddling up a stream, the speed of the boat (light wave) is slower to a guy on the shore (Earth).

That way, it is claimed that galaxies seen in this photograph are not in fact, 48 billion light years away but are, instead, merely 13.7 billion light years away. Just like the galaxies in the HUDF. That's 34.3 billion light-years' worth of galaxies in a thin expanding globe surrounding Earth 13.7 billion light years away Copernicus style.

The world of science now hinged on the assumptions of Copernicun gravity, flying photons, Cavendish Earth, gravity equality, 6.67, strange gases, random evolution, Darwin's timeline, neutral gravity, Helmholtz heat, constant light speed, Lorentz compression, flying electrons, nothing space, orbital electrons, Copernicun mass, floating continents, constant planet size, Eddington limitocity, omnipotent nothing, infinite photons, stellar fusion, Russian Expansion, doppelmatter, cosmic hatching, none of this exists, geocentric creation, salvation by ozone, white holes, dark matter, EPR, Higgs boson, microwave bang, CFC death, Hawking holes, omnibang, crazy ozone, dark energy, magical CMB, radio-birth, more dark matter, physics is optional, universal warpage, and that subspace is exploding near a galaxy far, far away.

This strange explanation does not address that if those galaxies are flying away from us, why are they emitting the same frequencies as local galaxies? If those galaxies are so much older, why are they the same colors and structures, even the same diversity as local galaxies?

There are several absurd implications related to this theory.

One implication of the 'stretching space-time theory' is that the region being viewed is actually $1/30^{th}$ of the size that it appears to be.

Am I to believe that those galaxies in the XDF are 1/30th the size of local galaxies? They are the same shapes, same colors, and same distribution of same galaxy types as galaxies which exist locally.

That violates a whole host of logic and law, so let's try out a variant logic: The galaxies are of standard size, but the space between them is 1/30th the size it appears to be.

Oh, boy.

If the 'galaxies-not-crashing' paradox was bad before – that whole 'dark energy' thing – then it just got (...let's see... $f=m/d2... 30^2$) about 900 times worse.

Why would I possibly believe this 'Honey I Shrunk the Universe' version of expanding universe theory?

What is there left to believe when scientists are <u>custom-designing how data are interpreted</u> in order to suit some pre-disposed belief?

And what about that microwave background that emanates so powerfully out of every galaxy? What would WMAP look like if you excluded all the XDF-distance galaxies from the WMAP?

In 2012 Kathryn Hansen interviewed Pawan Bhartia concerning the discovery of the ozone hole and the banning of CFCs and concluded her article by saying, "Changes in the ozone hole now are not significantly driven by changes in CFCs, but instead driven by year-to-year changes in weather in the stratosphere."

CMB and Distant Galaxies

Far in the background of the HDF, the HUDF existed. Far in the background of the HUDF, the XDF existed. And in the background of the XDF, there is more again.

At 56x zoom, the background of the XDF emerges.

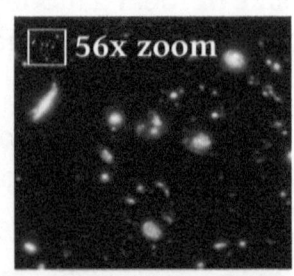

At 975x zoom, the background breaks up into distinct formations. Those are galaxies back there.

A conservative, superficial assay, shown below, indicates that there are over 800 galax-

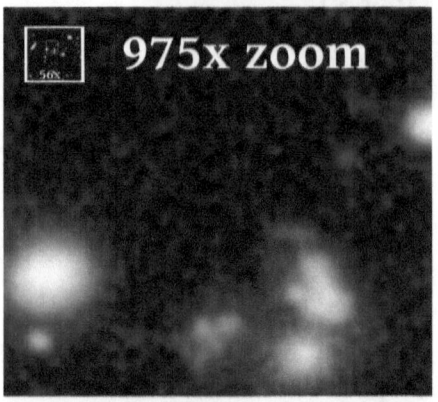

ies in this section of 1/975th of the Hubble eXtreme Deep Field. Each distinct color is used 100 times, 8 different colors were used.

A deeper analysis would find at least twice that many.

At 800 galaxies per section, the Hubble eXtreme Deep Field photograph shows 779,076 galaxies or 779 trillion stars.

When valiant, dedicated researchers and computers carefully crafted WMAP did they truly exclude all of the galaxies in the universe from that representation?

No.

We see that this universe a long, long time ago at a galaxy 48 billion light-years away existed as a place just like the cosmos around us today.

As for the CMB, to effectively map space that far away would require a 500-year dedicated Hubble Telescope mission. That's how long it would take to find enough galaxies for scientists to accurately remove galactic radiation from the cosmic microwave background radiation maps.

I do not recommend funding that project. I've already figured out what it would look like.

by James O Harris

It may be true that there is not a single point in the sky not occupied, at some distance, by a galaxy.

That's a lot of galaxies. How many are there?

The Universe: To Infinity or Something

How big is the universe?

In 1924, Hubble started with 32 galaxies.
In 1975 there were 15,000 galaxies.
In 1995, there were 100,000,000,000 galaxies.
In 2016 there were 2,000,000,000,000 galaxies.
In 2019 there are 400,000,000,000,000 galaxies.
Including the XDF background galaxies, 56,660,000,000,000,000 (57 quadrillion) galaxies.

This trend is expected to continue ad infinitum. The deeper

you look, the more you will see.

Is the universe infinite?

I must say that the universe is not infinite.

I will not predicate any prediction on any infinity or dis-infinity (singularity). I never will.

I cannot declare that the universe is infinite until we have measured it all, and, when we have measured it all, that we have added it all up, and when we have added it all up, that we have counted to infinity. Then and only then will I agree that the universe is infinite.

Yes. That is a paradox. We can't count to infinity. Therefore, my suggestion that I *might* agree that the universe is infinite is a lie. It's not true.

The key to this paradox is that all infinities are paradoxes. All infinities, dis-infinities, and singularities are paradoxical. You can't count that high, you can't count that low, and even if you could, the associated theory would not yield any useful information, testable prediction, or valid result.

Paradox Null
~~Infinity 1/0~~
~~Dis-infinity 0/1~~
~~Singularity~~

The proper way to approach theoretical numbers like the energy of the universe, the size of the universe, and the mass of the universe is to refer to them as *undefined.* Not infinite.

The proper mathematical symbol to describe a theoretical quantity is ">" (*greater than*) or "<" (*less than*).

Within references which formerly employed the word 'infinite', the term 'greater than' is is the proper replacement. Since the number found or calculated cannot refer to the whole universe, the value of the universe is greater than (>) or *more than* the number referred to.

The universe has more than 400 trillion galaxies.
The age of the universe is greater than 13.7 billion years.
The radius of the universe is greater than 1E+10parsecs.

A complaint which a theoretical mathematician might levy is, "We have worked out a system for working with infinities. We treat them as one would any other algebraic variable, we just never quantify it. We object because the terms *greater than* and *less than* would make these equations irrational."

Buddy, those equations are already irrational.

Concerning references of the infinitesimally small, the term 'less than' or 'null' may be appropriate. When numerical values are not readily apparent, a logical argument may be put in place of a number.

The probability of a single intelligent being creating the whole Earth by opening his mouth is less than the probability of the universe occurring from a big bang.

Thus we value one infinitesimal relative to another infinitesimal and successfully accomplish dis-infinities without any mathematical naval-staring.

The probability of Earth being created 6 days ago is
less than the probability of hydrogen spontaneously condensing into a fusion-bearing star.

On November 10, 2015, a team of 22 physicists at the University of Vienna published results of an experiment where they set up reception devices 143 kilometers apart, on the Canary Islands of La Palma and Tenerife, and recorded 'quantum entanglement'.

They even went so far as to avoid 'observing' it every way possible, incorporating random number generators in the computer program they used so that nobody knew when testing was actually happening. Not even the computers.

No matter how much they observed it or didn't observe it, that light was entangled even at that distance. No uncertainty about it.

The trouble of this observation is that it does not agree with either Heisenberg uncertainty, quantum entanglement. Uncertainty would have spun flying photons differently. Dirac's quantum entanglement was the product of observation, and that wasn't happening. The only 'flying photon' model this supports is spooky action.

The world of science now hinged on the assumptions of Copernicun gravity, flying photons, Cavendish Earth, gravity equality, 6.67, strange gases, random evolution, Darwin's timeline, neutral gravity, Helmholtz heat, constant light speed, Lorentz compression, flying electrons, nothing space, orbital electrons, Copernicun mass, floating continents, constant planet size, Eddington limitocity, omnipotent nothing, infinite photons, stellar fusion, Russian Expansion, doppelmatter, cosmic hatching, none of this exists, geocentric creation, salvation by ozone, white holes, dark matter, EPR, Higgs boson, microwave bang, CFC death, Hawking holes, omnibang, crazy ozone, dark energy, magical CMB, radio-birth, more dark matter, physics is optional, universal warpage, exploding subspace, and that photons magically affect other photons whether observed or not.

In 2016, a team including J. T. Nielsen, A. Guffanti & S. Sarkar evaluated the supernova data including the data connected to the 2011 Nobel Prize for Physics. They determined that no acceleration was occurring. They reported their statistical analysis as demonstrating a uniform rate of expansion, and pointedly expressed that data may be skewed based on the method of assigning red-shift values to various objects.

The reports of acceleration of expansion based on supernovae data, some claim, were based on biased, falsified, cherry-picked data, and both reports, some claim, were a fraud.

Either way, the 'accelerating universal expansion' concept by the High-Z Supernova team and the Supernova Cosmology Project had been canonized and continued as a hot topic of discussion.

And since universal expansion acceleration is canonized, everyone asks, "What is causing or driving that?" The standard answer is, "More dark energy."

Dark Matter and Dark Energy Paradoxes

At this point in history dark energy have become even more paradoxical. Not only was there ANTIGRAVITY (that term makes physicists cringe) in the form of dark energy, there was, according to the supernova guys, an apparently infinite source of it, occurring from no apparent source, and occurring at an increasing rate.

That violates the basic physical principle of 'you can't get something for nothing' or, in thermodynamics language, 'input equals output'. Dark energy theory is output (anti-gravity) without any input.

Likewise, the theoretical substance 'dark matter' violates the input/output principle. Dark matter has gravity – the effect of having mass – but no momentum – no experience of having mass.

In other words, dark matter affects objects without being affected. It is like a binary orbit involving only one object.

Of course, this makes no sense. It is not possible. Dark matter and dark energy theory are both paradoxical. "Does this paradoxical stuff actually exist?" "Are there real, underlying physical laws which seem to violate everything we know?"

These and other questions were answered in October and November of 2016.

Anti-Hydrogen

In 2013, the Atomic Spectroscopy And Collisions Using Slow Antiprotons (ASACUSA) experiment at CERN produced the first stream of anti-hydrogen.

Focusing this stream of anti-matter into a chamber of helium resulted in mostly fission but briefly, possibly, produced anti-helium where anti-hydrogen combined with neutrons to very briefly form anti-helium.

The author posits that the reason for the positive (+) bias of matter – all protons are positive (+) – is that the structure of the aether-pi. Like hydrogen atoms, aether-pi have a positive-bias 'heavy' core and a transient, variable negative pole. This governs the bias of mass, gravity, and matter. Negative-bias matter is inherently unstable.

Anti-hydrogen is the only known, proven anti-matter ever produced. To generate anti-hydrogen, hydrogen and/or helium atoms are placed in a powerful magnetic field such as an anti-matter detector. The powerful magnetic field there inverts protons into anti-protons – protons with a negative-charge surfaces - or causes fission whereby helium atoms break apart and, during that fission, anti-proton(s) may briefly form.

When a proton is converted into an anti-proton and the charge is inverted, the aether-pi field induced by that atom is also inverted. This does not affect the luminosity of the atom. Since photons are either positive or negative, this has no effect on the transmission of light. Electrons, however, form in opposition to the field, so electrons are inverted, forming *positrons.*

The rate and frequency of positron induction remains identical to the rate of electron induction, so the spectral signatures of anti-matter are identical to the spectral signatures of matter.

The bias of the anti-proton does not change the submagnetic potential of the anti-proton, so the quantity of aether-pi attracted to that structure remains unchanged. This means that the mass and gravity of anti-matter is equal to the mass and gravity of matter.

Antimatter does not exist in nature because protons, unlike electrons, are permanent persistent particles. Therefore anti-protons, unlike electrons, won't just blink out of existence if the aether is disturbed and pop up somewhere else. Instead, because the aether-pi field is inverted, anti-protons experience gravity to protons at the rate of gravity to about the third power, or gravity cubed. Furthermore, there is no repulsion, as exists between protons of the same charge, so anti-protons immediately crash into the nearest proton, resulting in mutual fission.

When fission occurs the aether-pi scatter and a gravity wave is emitted as the disturbed aether-pi field renormalizes. Gravity waves are transmitted across the same medium as light waves, so gravity waves probably travel at the same speed.

Anti-Matter: Weapons of Uselessness

As seen in the investigations of fission, when matter is destroyed and its mass ceases, no net energy is emitted. When matter collides with antimatter, that output will be uneventful. It will be uneventful because antimatter larger than hydrogen nuclei do not exist and anti-hydrogen has very little energy potential.

At best, maybe a super-magnet could be developed from solid

anti-hydrogen.

Paradoxes of Anti-Bosonic Matter

In big bang theory, the universe sprang out of a 'nothing'-sized singularity. That singularity was too small for anything to physically exist in. The singularity had (little or) no space, so there could not be any matter there.

In the cosmic egg/singularity version of events, there was no *bosonic matter* – solid, atomic, persistent stuff – before a big bang.

So, where did all this *stuff* come from?

You can't get something for nothing, right? So to account for the abundance of stuff and to provide balance to the equation of the force, big bang speculators suppose that during the interaction of nothing with nothing that both matter and anti-bosonic matter sprang out.

This is the theory that for everything to exist, the opposite of everything must also exist in a literal, anti-physical sense.

The anti-bosonic universe would be the 'other half of the charge' of the singularity. It might be thought of a mirror universe, but that makes no sense since it cannot overlap or co-exist. If it did, it would inherently destroy both places. That's the definition of its existence.

So all the anti-bosonic matter flew away and went into hiding at the edge of the universe. We know this because we cannot find any.

Anyhow, the mathematical model of the e=mc2 singularity provided that if you combined all the matter and anti-bosonic matter back together, you would have no matter again and you could have a zero-size singularity.

One obvious flaw, here, is that since gravity and mass **are a**

single field which ceases when matter fissions, there exists no gravity or attraction after fission occurs. Then there is no gravity to draw anything together into a singularity.

Initially, the discovery of anti-particles was heralded as another proof of Einstein, general relativity, and big bang theory. But the anti-matter that physicists have locked up in super-powerful magnet fields is not as 'anti-matter' as some people thought it would be.

Particles and anti-particles and not fundamental opposites of each other. They are only electromagnetic opposites. Would anti-bosonic matter have the opposite of standard bosonic polarity or, since that is a material attribute, would it have the opposite of magnetic interaction?

Anti-particles have the same mass as their counterparts. They are the not literal opposites. Anti-bosonic matter should have 'negative mass' or the opposite of mass. It should also have negative gravity or the opposite of gravity.

If matter transmits light, anti-bosonic matter should do the opposite of that, too. That could mean any number of things: no emissions; no spectra; or no interaction whatsoever. But anti-hydrogen has identical spectra to ordinary hydrogen.

So anti-matter is <u>ordinary bosonic matter</u> which has been electromagnetically inverted. Anti-bosonic matter has never been discovered.

Anti-particle Detector Paradox

Since matter can be electromagnetically inverted to form anti-matter in the vicinity of a powerful electromagnetic field, and since all anti-particle detector technology involves very powerful magnetic fields, anti-particle detectors generate the anti-matter which they then detect.

Case in point: Scientists hoped to detect higher levels of anti-matter via satellite beyond Earth's atmosphere. Instead, they found less than they typically find on Earth's surface. The found the highest abundance of anti-matter in the aurora borealis.

There were not 'finding' anti-matter. They were making it. Earth's magnetic field aids anti-matter production on the surface. Space is fairly neutral

The borealis are caused by the electrical induction of winds blowing as Earth's polar jets. The polar jets launch gases beyond the reach of Earth's gravity and more than 50,000 kilometers high over the north and south poles. Elements heavier than hydrogen tend to fall out of that wind and, as they fall back to Earth, that motion *induces* static electricity which lights the sky.

Static electricity aids anti-matter generators and causes other nuclear reformation as well, generating carbon-14 and other interesting isotopes. The aurorae are like giant mass spectrometers in the sky electrifying atmospheric gases.

Aurora activity spikes when the energies of solar flares reach Earth.

When a satellite sees a 'spike in anti-matter particles' it is detecting one of two things: Increase in the electromagnetic flux (as per a solar flare); or an increase in the density of the solar wind. Either of these events will increase the production efficiency of an anti-matter generator.

2016 - The Perfect Scientific Storm

Three independent scientific experiments took place in October and November of 2016.

These three famous-until-they-happened experiments stirred up more fanfare than the scientific community had seen since the first atom bomb exploded. Headlines around the world sang in chorus, "Dark Matter and Dark Energy About to be Discovered."

These three major experiments involved, collectively, 5 supercomputers, 4 anti-matter detectors, 3 satellites, 2 million dollars, CERN's atom-smasher, and a dozen valid PhDs.

They were all looking for the same things: Dark matter and dark energy.

They all found the same thing.

Nothing.

Three different high-energy, high-probability experiments conclusively found that dark matter and dark energy *do not exist* "as predicted".

The "as predicted" part might be an attempt at persuading lawmakers and donors to continue their funding. The fact is that those three studies all completely disproved the dark theories. The theoretical substances did not exist *as predicted*. In other words, *if* dark matter or energy existed as predicted, *then* the experiments would have found them.

That did not happen.

Now, with those findings in mind, if dark matter or energy *do* exist, *then* those substances do not have the properties of dark matter or dark energy such as they were designed to have.

So at this point in history, if dark energy exists, it does not have the property of anti-gravitation. If dark matter exists, it does not posses the attributes of either mass or gravity.

So as of 2017, if dark matter and dark energy exist, they do not exist within the capacity or structure of their design. That means that dark matter, if it exists, will not resolve the galaxy spin riddle, and dark energy, if it exists, will not solve the gal-axies-not-crashing riddle.

This lands us back in the middle of the grand galaxy paradox:

If Newton's gravity is true, then how is Newton's gravity true?

Somehow it is.

In a speech at Trinity College, Cambridge University, published by the Royal Institution on February 15, 2017, and on the heels of the massive failures in the fields of dark matter and dark energy, David Tong said, "If the LHC (Large Hadron Collider) doesn't see something within, say, a two year time scale, it seems very, very unlikely that it's going to see something moving forward."

Two years later, no new discovery had taken place. No new breakthrough achieved.

Not in Higgs boson theory.
Not in God particle theory.
Not in mass theory.
Nothing.

The world of science now hinged on the assumptions of Copernicun gravity, flying photons, Cavendish Earth, gravity equality, 6.67, strange gases, random evolution, Darwin's timeline, neutral gravity, Helmholtz heat, constant light speed, Lorentz compression, flying electrons, nothing space, orbital electrons, Copernicun mass, floating continents, constant planet size, Eddington limitocity, omnipotent nothing, infinite photons, stellar fusion, Russian Expansion, doppelmatter, cosmic hatching, none of this exists, geocentric creation, salvation by ozone, white holes, dark matter, EPR, Higgs boson, microwave bang, CFC death, Hawking holes, omnibang, crazy ozone, dark energy, magical CMB, radio-birth, more dark matter, physics is optional, universal warpage, exploding subspace, total entanglement, and that dark theories can't be wrong even though they are.

On April 15, 2019 *Nature* reported that the life-span of a neutron is 14 minutes, 39 seconds on *average*. Or 14 minutes, 47 seconds when using a different method.

The American Physical Society met on April 13 and 14 of 2019 in Denver, Colorado, where this conundrum was a hot topic.

The first method is the *bottle method* where a quantity of neutrons are 'bottled' and the population of neutrons is periodically measured. The *beam method* takes a different approach: a beam of neutrons is generated, and the resultant protons are counted to indicate how long the neutrons existed.

The trouble here, I believe, is that these are two different measurements entirely. As seen through wave-particle transformation, the interactions and potential outcomes are many. It seems that, based on these two differing measurements, that 0.9% of neutrons do not become protons. They do not remain neutrons, either. This suggests that 0.9% of neutrons do not survive the expulsion of the electron field, but experience fission instead.

The bottle method counts how many neutrons remain. This indicates that *unstable neutrons* are the first to go, making the bottle method count deterioration *as including* neutrons that no longer exist at all.

The beam method ignores unstable neutrons, counting only the stable neutrons which successfully transformed into protons.

PART II:

PARADOX NULL

2019 – BREAKING OUT

Here is the solution to each of those paradoxes in short form.

~~Copernicun gravity~~ Low-mass Earth
~~flying photons~~ Light-waves
~~Cavendish Earth~~ Low-mass Earth
~~gravity equality~~ Low-mass Earth
~~6.67E-11~~ 2.76E-10
~~strange gases~~ 2.276E-10; low-mass Earth
~~random evolution~~ Changing environment
~~Darwin's timeline~~ Natural selection
~~neutral gravity~~ Incomplete theory
~~Helmholtz heat~~ Atomic Energy Transformation (AET)
~~constant light speed~~ Variable light speed
~~Lorentz compression~~ Variable light speed
~~flying electrons~~ Wave-Particle Transformation (WPT)
~~nothing space~~ Aether-pi space
~~orbital electrons~~ Wave-Particle Transformation (WPT)
~~Copernicun mass~~ External Mass
~~floating continents~~ Radial expansion
~~constant planet size~~ False
~~Eddington limitocity~~ Atomic Energy Transformation (AET)
~~omnipotent nothing~~ Aether-pi space
~~infinite photons~~ Light-waves
~~stellar fusion~~ Ordinary planets; AET
~~Russian Expansion~~ The Law of Time, AET
~~doppelmatter~~ False

~~the cosmic hatching~~ False
~~none of this exists~~ False. See Wave-Particle Transformation (WPT)
~~geocentric creation~~ False
~~salvation by ozone~~ False
~~white holes~~ Local nucleogenesis
~~dark matter~~ Low-mass black holes, 2.76E-10
~~EPR~~ Wave-Particle Transformation
~~Higgs boson~~ Aether-pi space
~~microwave bang~~ Ancient universe
~~CFC death~~ False
~~Hawking holes~~ Ordinary planets
~~omnibang~~ False. See local nucleogenesis
~~crazy ozone~~ False
~~dark energy~~ False. See low-mass black holes.
~~magical CMB~~ Ancient universe
~~radio-birth~~ Continual nucleogenesis
~~more dark matter~~ Low-mass black holes, 2.76E-10
~~physics is optional~~ Physics was incomplete
~~universal warpage~~ variable light speed
~~exploding subspace~~ Ancient universe
~~total entanglement~~ Light is transmitted as waves
~~really dark theories~~ Low-mass black holes, 2.76E-10

That is a list of just 46 of the paradoxes that this book addresses and eliminates. This list is somewhat generalized in that I could detail not less than six paradoxes as relate to stellar fusion theory, ten relating to Cavendish Earth, and fifteen relating to big bang theory.

I could have easily doubled this list. I also could have filled this book with equations.

Resolving the Paradoxes of Galaxies

According to the infamous dark experiments, there is no dark matter or dark energy. Neither of these unseen forces or sub-stances can exist within the solution to these paradoxes.

Constraints to solving this riddle include: We must assume Newtonian gravity to be true and correct; we must assume that observations of S2*'s orbit are accurate.

Constraints we are not bound by include: General Relativity, Special Relativity, MONDS (Modified Newtonian Dynamics), and Quantum Mechanics. Those theories have all been failing at this for more than 35 years. Some have been failing for 100 years now.

And if we find what dark energy and dark matter theory sought for, then we will find a solution where no paradox is implied or exists in any fashion.

The solutions to the riddles of galaxies requires redefining time.

FOUR PRINCIPLES OF THE FOURTH DIMENSION

Before getting into how the *Law of Time* resolves the galaxy paradoxes, there are four fundamental principles of time. These principles are important to understanding what time is and what it means when the speed of time changes. These principles also reveal why changes in the speed of time are hard to detect.

The First Principle of the Fourth Dimension

The speed of light is 299,792,458 meters per second.

This measurement is a unit of time. The meter referred to is counted as a solid, unchanging material unit.

On Earth, the speed of light is 299,792,458 meters per second. What happens if the speed of time on Earth doubles?

If the speed of time doubles, then the speed of light doubles. However, the period of time during which one second transpires is reduced to one half. To calculated the new speed of light, we may multiply the speed light by two and divide the period of time by two.

$$(299,792,458 \text{ meters} * 2) * (1 \text{ second} / 2) =$$

299,792,458 meters per second.

This simple math establishes the First Principle of the Fourth Dimension: The speed of light – *as locally experienced* – remains constant no matter what the speed of time is.

Harris Blue-Shift

The frequency of light is variable.

Because the speed of light and time are variable, the frequencies of light received from objects are not the same frequencies as the light which left those objects. Because the speed of time changes, and the speed of light changes with it, light is red-shifted and blue-shifted during transmission.

Harris Blue-Shift calculates time dilation based red-shifting and/or blue-shifting of light.

Harris Blue-Shift works for everything. Stars, planets, moons, black holes, distant galaxies... everything. Harris Blue-Shift is to be used in all cosmological calculations of frequency, wavelength, and temperature.

The equation for Harris Blue-Shift is:

Hz received = Hz emitted * (speed of light emitted / speed of light received)

Using terminology from the Law of Time, the equation looks like this:

Hz emitted * TCF of object observed = Hz received

The term TCF is short for *Time Correction Factor*. An object's TCF is how fast time occurs for that object proportional to how fast Earth experiences time. Earth has a TCF of 1.0.

The Second Principle of the Fourth Dimension

**The speed of all electromagnetic and inertial events
are proportional to the speed of light.**

As with light speed calculations, all calculation involving torque, watts, Newtons, and so on are proportional to the speed of time. If the speed of time on Earth doubled, gravity would double, but the length of one second would get cut in half, and the gravity experienced would remain the same.

$$(9.8 \text{ Newtons} * 2) * (1 \text{ second} / 2) = 9.8 \text{ Newtons}$$

All subatomic, atomic, and macro-atomic inertial and energetic events occur proportional to the speed of time.

The Third Principle of the Fourth Dimension

**Temperature is an electromagnetic event
proportional to the speed of time.**

The temperature of an object involves energetic interactions and is proportional to time. This means that when viewing a stellar object which experiences a different speed of time, the temperature which that object experiences locally may be very different than the apparent temperature.

The Sun is viewed as having a surface temperature of 5,778 Kelvin. The TCF of the Sun is 9.39. To find the temperature of the Sun as locally experienced, divide the apparent temperature by the the TCF of the Sun.

$$5,778° \text{ Kelvin} / \text{TCF of } 9.39 = 615.48° \text{ Kelvin}$$

The surface temperature of the Sun, as locally experienced, is about 615.48° Kelvin or 648° Fahrenheit.

The Fourth Principle of the Fourth Dimension

Relative motion does not change any of these rules.

Relative motion is motion against an *inertial frame of reference.* The speed of light is constant within an inertial frame of reference. Michelson, Morley, and Einstein believed that the speed of light was universally constant, therefore, <u>Michelson, Morley, and Einstein thought that the whole universe was **one singular inertial frame of reference.**</u>

It's not.

Inertial Frame of Reference

The inertial frame of reference of any atomic or macro-atomic structure is an active, dynamic field consisting of unique, individual particles which constitute a medium.

This constitutes a new aether theory, and since this aether theory has many new, unique attributes not identified in pre-existing aether theories, this aether must have a unique name. It is called aether-pi.

Aether-pi

It is referred to as aether-pi, in part, because the density of the aether-pi are calculated using the mathematical term pi. Since the behavior of these gaseous-state aether-pi is atmospheric relative to atomic nuclei, calculations of aether-pi involve circles, spheres, and the term pi.

That is absolutely different from former aether theories. Primative aether theories were founded on the idea that the 'Lumineferous Aether' or 'space-time' was universally homogenous (same everywhere), isotropic (same temperature/density everywhere), static, immobile, and permanent.

Formerly, the inertial frame of reference was thought to consist of nothing. It was just a location. It was an arbitrary numerical descriptor of where an object existed.

Formerly, the inertial frame of reference was not a thing. It did not actually exist.

How fast does the inertial frame of reference occur? At the speed of light. Energies and events across an inertial frame of reference are transmitted at the speed of light. That is why *the speed of light* is *the speed of light*.

A Universe Divided

The inertial frame of reference is localized to any given object at the speed of light. Because of this, relative motion cannot be compounded. If I accelerate to 600 miles per hour, the inertial frame of reference, like the air inside of a cockpit, is not traveling at all. Whether I am traveling zero miles per hour, 600 miles per hour, or 6 million miles per hour does not matter. Unless I am experiencing acceleration, I am not experiencing 'relative motion'.

Therefore, relative motion, as defined by Einstein, is fictional and the effects of relative motion described by Einstein only occur during acceleration.

THE LAW OF TIME

The Law of Time is a new Newtonian Law which dictates that the speed of time is proportional to mass proximity. It is expressed as the following equation.

$$time = mass / distance^{(1/2)}$$

For two or more bodies:

$$time = (mass_1 / distance_1^{(1/2)}) + (mass_2 / distance_2^{(1/2)}) + (mass_3 / distance_3^{(1/2)}) + \ldots n$$

Time is measured in kilograms/meter^2.

Time is expressed in terms of weight.

The overall effect of this law is that the speed of time is positively associated with mass proximity – the closer you are to more stuff, the faster time happens.

Motion-related and/or gravitational time dilation may be considered when calculating local time speed.

The Speed of Time

The density of the aether-pi field is equal to the speed of time.

$$aether\text{-}pi\ density = time = mass / distance^{(1/2)}$$

NOTE ON RELATIVE MOTION TIME DILATION

When calculating time dilation due to relative motion, the speed of motion should be calculated relative to the motion against Earth's atmosphere when on the surface of Earth. Above 25,000 feet, the Sun and Earth have nearly equal motion-related influence. When beyond the orbit of the moon, relative motion should be calculated relative to the Sun or other objects of influence. When not in vicinity of massive objects, the speed of motion should be calculated relative to the object in question.

A very important point: Relative motion is primarily a 'skin effect'. It is something that happens on the surface of an object. It happens where the mass or gravity field of that object experiences motion relative to the *ambient* gravity field.

Since that field is localized atomically, relative motion does not occur, for instance, in the middle isle of an airplane. The window, however, can experience significant relative motion.

If you ride in an airplane with one hand on the window and the other hand towards the middle of the plane, your hand on the window will experience less time than the other hand.

If you go skydiving, the skin on your face will experience time more slowly than your brain does.

The relative motion time dilation which occurs is due to the *mixing* of the mass/gravity fields of the various objects involved. When an airplane flies through the atmosphere or a skydiver falls through it, the mass/gravity field of two objects mingle. Aether-pi are traded from one field to another. These trades inevitably result in inefficiencies. Those inefficiencies are the lowering of the density of the gravity/mass field. That density is directly proportional to the speed of time. Thus, an object in motion <u>within the gravitational field</u> of another object will experience *time dilation*.

Einstein-model time dilation is not accurate. It involves the speed of light. Nothing ever moves with respect to the local speed of light except <u>during acceleration.</u> Einstein's model of time dilation is an equation regarding *acceleration* instead of motion and *only applies locally.*

Einstein's time dilation equation

$$t_d = t \ / (1 - \text{velocity}^2 \ / \text{ speed of light}^2)^{(1/2)}$$
$$t_d = t \ / (1 - v^2 / c^2)^{(1/2)}$$

May be rewritten

$$t_d = t \ / (1 - (\text{acceleration}^2 \ / \text{ speed of light}^2)^{(1/2)}$$
$$t_d = t \ / (1 - a^2 / c^2)^{(1/2)}$$

When an object is accelerated almost instantaneously to the speed of light, the mass of that object becomes zero. It moves faster than the its inertial frame of reference – faster than gravity, mass, and magnetism. This is how nuclear bombs get all that extra bang – by accelerating matter to fractions of luminal speeds almost instantaneously. That results in a loss of mass, a loss of the strong nuclear force, and results in fission.

The inertial frame of reference also includes *time.* When an object is accelerated beyond the speed of light and loses its inertial reference frame, it ceases experiencing time. That is when fission occurs.

INTRODUCTION TO THE AETHER-PI

The aether of the author is an original breed of aether in at least several respects, so it is named aether-pi. It is called aether-pi because it's presence is atmospheric – spherical. Calculating that involves pi. But, mostly, I'll choose that term because it is fun and memorable.

A fundamental attribute embedded within the atmospheric nature of the aether-pi is this: aether-pi are dynamic. They move with the atom. They exchange between atoms.

This is greatly different than other aether theories in a least one major way: <u>Aether-pi do not have any permanent intergalactic address.</u> Michelson-Morley's Luminiferous Aether. Lorentz's Aether, and Einstein's space-time were all universally fixed, static entities. Aether-pi are mobile and highly dynamic. When the Earth moves, the aether-pi move with it.

Aether-pi, like atmospheric gases, are not intrinsically attached to any atom or object.

One thing aether-pi *do* have in common with Michelson-Morley's LE is that the aether-pi conduct light as waves. Aether-pi are the medium of transmission.

Since aether-pi are localized to objects individually, and aether-pi conduct light waves, <u>the speed of light near any object is constant to **that** object.</u>

There is no aether flow.

Since aether-pi are dynamic, localized to objects individually, and are also the medium which conducts light, the speed of light is <u>locally</u> constant no matter how you measure it, when you measure it, or what you measure it with.

NOTE ON NEWTON'S FIRST LAW OF MOTION

Newton's first law (popular modern version) goes: In an inertial frame of reference, an object either remains at rest or continues to move at a constant velocity, unless acted upon by a force.

First term: Inertial reference frame: The aether-pi of an individual object are, in total, the inertial reference frame of that object.

Force: Force is any imbalance in the inertial reference frame including magnetism and gravity. Note: Magnetism as atomic repulsion is the mechanism whereby momentum is transferred from one object to another. All interactions described as inertia, momentum, and force are magnetic interactions.

So I simplify things:

First Law of New Newtonian Motion:

> **No object experiences acceleration or deceleration unless acted upon by force.**

There are several reasons why this may be pared down so much. In Michelson-Morley/Einsteinland, everything was always in motion compared to some universal standard. Thus, nothing was at rest. Ever.

But something *could* be at rest, so in Einsteinland there are three states of motion: Motion, rest, and acceleration.

But with the dynamic aether-pi, there are only two states of motion: acceleration and rest. There is no such thing as 'motion' as locally experienced.

Note: This the law does not limit or imply the quantity of force(s) in play.

NOTE ON NEWTON'S SECOND LAW OF MOTION

'In an inertial frame of reference, the vector sum of the forces (F) on an object is equal to the mass (m) of that object multiplied by the acceleration (a) of the object. (It is assumed here that the mass is constant – see relativistic mass.)'

Once again, we can remove the inertial frame of reference caveat. The rest of the text is fine, but the parentheses gotta go. Relativistic mass is a myth. The mass of an object is a constant value. It is the quantity of the aether-pi as the atmosphere-like field surrounding an atomic nucleus, a macro-atomic structure, and/or any object.

The Second Law of Motion may be simplified.

The acceleration of any object is equal to its mass divided by force applied.

Atomically, the totality of the aether-pi field is proportional to the submagnetic attraction of the atomic nucleus. The atomic nucleus does not change – and certainly does not exponentially increase in charge or size – when in motion. As related concerning Newton's First Law of Motion, mass is a constant value.

The equations stands on its own better than before.

Force = mass * acceleration

THE REVOLUTION OF MASS

Many keen physicists caught on to what just happened in the section above. For the rest of us, I'll say it clearly:

The mass of an object is external to that object.

Everyone since Isaac Newton knew that gravity was external to an object. According to Isaac Newton's equations, mass is proportional to gravity. It should be of little surprise, then, that gravity and mass occur in the same place.

This is utterly revolutionary because major atom smashers – from the beginning of the history of major atom smashers – were designed with the purpose of finding mass. Where were they looking for mass? After the Einsteinian tradition of empty nothing space and following the concepts posited by Rutherford and Eddington, scientists were looking for mass *inside* of the atom.

The nucleus of an atom contains about 10% as many (solid-form) aether-pi as the aether-pi which exist as the (gas-form) field induced by that nucleus. <u>The aether-pi of the nucleus have no direct interaction with other atomic nuclei.</u> Since mass and gravity are defined by extra-atomic effects, the nucleus of an atom has no mass. All mass exists external to the nucleus.

This leads to new definitions of mass and matter.

The nucleus of an atom is *matter.* The field surrounding the nucleus is *mass.*

Photons and electrons (PEs) do not have any mass. PEs only occur *within the mass* of an atom. Photons and electrons are secondary attributes of mass. Mass is a secondary attribute of matter.

Are photons and electrons matter? That's debatable. Do they have mass? No. Aside from the logic above, photons and electrons clearly do not have mass otherwise hot gases – gases with high photon populations – would be heavy and fall. Instead, hot gas atoms take up more space and weigh the same weight, so they become lighter – atomic distances increase, density falls, the proportion of mass to space decreases, and the gas becomes lighter.

Photons and electrons are particles, but unlike protons and neutrons they are *neutral* to the field they occur in. Particles which are neutral to the aether-pi field do not have mass. Since those particles do not reorient the field, those particles do not experience time, mass, gravity, or momentum.

Matter experiences all those things.

Matter is also centralized and persistent. Matter does not blink in and out of existence. Photons and electrons do not fit definition of matter *unless* they become nucleic or structural components as through photosynthesis.

Back at the atom smashers, it was confirmed there that electrons had no mass.

Although that violated what Einstein posited, that did not surprise scientists very much.

What did surprise them is that protons did not have any mass either. In fact, when they smashed protons, they not only found no mass, they also found that they got less energy out of that fission than the energy they put into causing that fission.

These experiments indicate that mass is not proportional to energy, and that mass cannot be converted into energy.

The 'missing energy' is expended through fission as gravity waves – disturbances and a slight increase to the density of the aether-pi field.

NOTE ON EINSTEIN'S RELATIVISTIC MASS

The inertial frame of reference of an object is the aether-pi field which interacts with that object. The linear interactive speed at which that field communicates energy is the speed of light. If an object approaches the speed of light relative to its own inertial frame of reference, that object will lose communication with its inertial frame of reference.

Before reaching the speed of light, the inertial frame of reference is completely lost. Time, magnetism, and the strong nuclear force cease and fission occurs.

That chain of events reveals three things to address concerning Einstein's concept of relative motion. Here is Einstein's equation.

$$\text{relativistic mass} = \text{mass} / (1 - \text{velocity}^2 / c^2)^{(1/2)}$$

First off, as speed increases, mass diminishes instead of increasing. To mathematically reflect that trend, swap the speed of light (c) with the velocity.

$$\text{mass} = \text{mass} / (1 - c^2 / \text{velocity}^2)^{(1/2)}$$

Now we have to address the fact that the aether-pi are active and mobile. Because of that there is no such thing as relative motion. If you do not feel pushed or pulled in any direction, you are not experiencing motion relative to your inertial frame of reference. Motion relative to the inertial frame of reference is experienced as *acceleration.*

To finish correcting Einstein's equation, replace *velocity (motion)* with *acceleration.* We will add a little pi symbol 'π' to indicate the *acceleration mass* or *aether-pi* mass of the object experiencing acceleration.

$$\text{acceleration mass}_\pi = \text{mass} / (1 - c^2 / \text{acceleration}^2)^{(1/2)}$$

$$m_\pi = m / (1 - c^2 / a^2)^{(1/2)}$$

In *Illustration Relativistic Mass versus Acceleration Mass*, an orange line represents Einstein's relativistic mass, a yellow line represents acceleration mass, and a blue line represents the speed of light. The chart stops at 300,000,000 meters per second or 100.07% of the speed of light.

At just about the speed of light, relativistic mass skyrockets. At 295 million meters per second relativistic mass climbs to 462%. At 299,792,457.9 meters per second relativistic mass climbs to 3,871,543%.

At 295 million meters per second acceleration mass falls to 17.8%. At 299,792,457.9 meters per second acceleration mass falls to 0.0000258%

Atomic and nuclear weapons function according to this law. Atomic detonations are successful when mass falls to around 85% which occurs at about 54% of the speed of light or 160 million meters per second.

A uranium fission chain reaction is sufficient to obtain these speeds, and is accomplished, in principle, by taking one cannon-ball rich in uranium 235, loading it into a cannon, and firing the first that cannon-ball into a mass of uranium 235.

In the case of a bomb or missile, an atomic explosion happen in four stages.
  1. The trigger
        • A small charge situated in the nose of the bomb

        explodes on impact,
- A radar-ranging scope activates bomb prior to impact,
- Or a computer-controlled detonation device initiates explosion.
- 500 meters per second.

2. The gunpowder
   - The trigger ignites a conventional explosive, sending a cannon-ball of 235U crashing into other 235U.
   - 5,000 kilometers per second.

3. First fission
   - The 235U goes critical, releasing barium, helium, and hydrogen as fission occurs.
   - 5,000,000 kilometers per second

4. Uranium, barium, helium, and hydrogen are then accelerated into collisions
   - These collisions begin at first fission speed.
   - When one atom strikes another, the fissile material of the second atom has greater linear speed than the material of the first atom.
   - This results in a compound cascade fission reaction.
   - This condition is supercritical.
   - <160,000,000 meters per second.

Yield is proportional to speed. Nuclear weapons involve higher speeds and lower masses. Nuclear bombs also happen in four stages They are designed differently to get to step 4 faster.

In the case of a bomb or missile, an atomic explosion happen in four stages.

1. The trigger
2. The gunpowder
   - The trigger ignites a conventional explosive, sending a cannon-ball of 235U crashing to-

wards other 235U.

- 5,000 kilometers per second.
- The cavity between the two units of 235U is full of hydrogen isotopes which the cannon-ball compacts to critical conditions.
- Fission begins
- < 190,000,000 meters per second.
- The cannon-ball reaches its destination
- more fission begins

3. First fission
   - The 235U goes critical, releasing barium, helium, and hydrogen as fission occurs.
   - Fission components exist already, initiating supercritical conditions
   - 50,000,000 kilometers per second

4. Uranium, barium, helium, and hydrogen are accelerated into collisions
   - When one atom strikes another, the fissile material of the second atom has greater linear speed than the material of the first atom.
   - Supercritical cascade fission reactions occur.
   - < 200,000,000 meters per second, mass below 25%
   - Five to ten times more material is destroyed is the fission of this bomb than in a comparable atomic bomb.

A hydrogen bomb has two supercritical phases whereas an atomic bomb only has one.

With this information, we can redraw the acceleration mass graph to include this natural limit. This natural limit varies for different substances. Uranium235, for instance, has an unusually low *acceleration fission limit (AFL)*.

The acceleration fission limits occurs a given atom loses a given percentage of its inertial frame of reference – perhaps

50% - at which point in time the strong nuclear force ceases to maintain the atomic nucleus. The nucleus then explodes like an atomic grenade (if it's large or dense enough) or fizzles away if it is a lighter element.

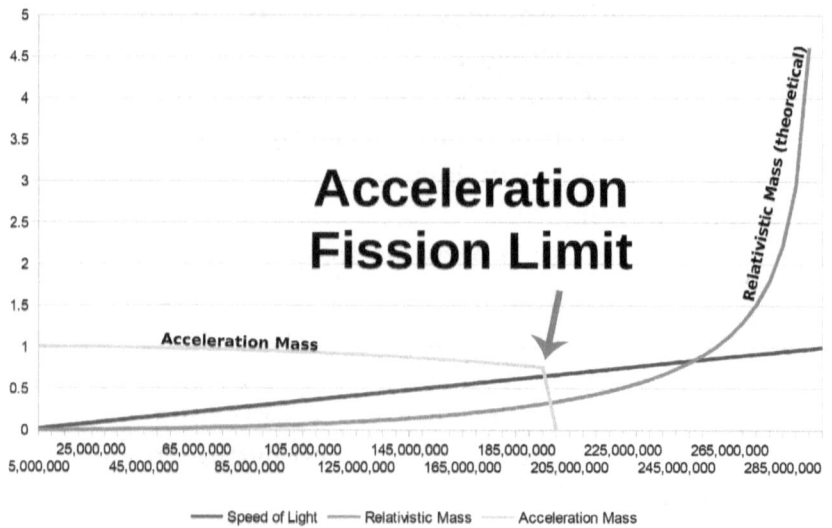

Essentially, when matter begins to fail, it explodes like a balloon with a hole poked in it. If enough of it explodes fast enough, a chain fission reaction occurs.

NOTE ON ENTROPY

Light waves consist of entropy. They are energetic disturb-ances within the aether-pi which are broadcast outwards at the speed of light. Those disturbances continue outwards at the speed of light until disturbed by material interaction. When disturbed by material interaction, the light energy is changed. It is deflected, reflected, transformed and, as con-cerning the initial light wave, entropied.

Matter entropies light.

But that is only half of the story. Matter also organizes light. Radium, for instance, organizes incoming energies to a steady output frequency of 9.2 billion cycles per second.

That leaves basically two ways to look at things. Either:

    1. All material interactions cause entropy. Or,
    2. All material interactions eliminate entropy.

In either case, all entropy is entropy of other entropy and input equals output.

$$input = output$$

So what happens if I try to convert mass to energy? Mass does not entropy.

Mass, which is also the field of gravity, does not diminish, re-duce, or go away. Mass is also the field in which disturbances occur as light-waves.

Mass is a field held in place by matter. Matter does not en-tropy.

Matter is the organizing principal, mass is the effect of matter, and light-waves are an attribute of mass.

Since matter is what causes mass, in order to accomplish mass/energy conversion matter must be diminished to diminish mass. Mass/energy conversion, then, is matter/energy conversion.

Energetic transmissions including light-waves are an attribute of matter.

Can matter be converted into energy? Can I convert an object into an *attribute of that object*?

Can I change a piece of white paper into 'white' or into 'paper', eliminating the other attribute?

No.

Therefore, neither gravity nor mass are accurately defined as electromagnetic like light-waves, photons, or electrons. Neither one relies on electrons. Neither one is made of electrons. Neither of them spreads. Neither travels at the speed of light. Neither one travels at all.

THE LAW OF TIME RESOLVES SAGITTARIUS A*

Using data which indicates the effect of Sagittarius A* on the speed of time locally, the mass of Sagittarius A* has been resolved as totaling 4.61E+34 kilograms.

That is equivalent to 26,446 solar masses.

That is a lot less than conventional estimates of 8.58E+36 kilograms or 4.31 million solar masses – merely 0.54% as much.

So how can such a little... well, such a *huge* mass but not 'supermassive mass' whip S2* around at 7,650 km/second?

It's not going that fast.

The TCF of Sagittarius A* is about 38,133. Sagittarius A*, weighing 26,446 solar masses has a TCF of 38,133 because of itself. Other objects in the vicinity also raise the rate of time, so the actual TCF of Sag A* is higher.

S2* does not experience time as fast as Sag A*. S2* has a TCF of 31.84. Once again, this does not include uncalculated local effects, but includes SagA* and S2*. This is a low or conservative estimate.

Based on that (low) estimate, S2* _as locally experienced_ does not travel 7,650km/s at perihelion. Instead, it is traveling 1/32nd

that fast. At maximum perihelion speed, it is traveling less than 240.3 kilometers per second. To calculate Sag A*'s mass from S2*'s orbit, you have to calculate it based on S2* traveling 240.3 kilometers per second at perihelion.

In other words, watching S2* orbit Sag A* in real time is like watching in 32-speed fast forward. If you base your calculations of Sag A*'s mass on the fast-forwarded version you are watching, your calculations will exponentially increase the mass of Sag A*.

To put S2*'s speed of 240.3 km/s into perspective, the Sun orbits Sagittarius A* at about 220 km/s and Mercury orbits the Sun at 47.3 km/s.

Resolving Dark Matter

Based on that newly calculated mass, Sagittarius A* has virtually no gravitational effect on the Sun – 0.54% as much gravity as formerly believed.

Sagittarius A* has very little gravity beyond the central galaxy.

Based on the mass of Sagittarius A*, there is no need to explain why the galaxy does not behave like a solar system orbiting Sag A*. The galaxy is not held together by Sagittarius A*. The outer galaxy, then, is held together by itself. Like a Frisbee. And like a Frisbee, it rotates all at one rate, where the speed of the outer parts is faster than the inner parts and the RPM (Rotations per Millennium) is the same for everyone.

The fact that Sagittarius A* has little mass removes the theoretical necessity for dark matter.

Though the Law of Time, the Milky Way galaxy exists as observed, Sagittarius A* exists as observed, S2* exists as observed, and there is no paradox. There is no dark matter.

Paradox Null

~~Galaxy Spins~~
~~Dark Matter~~

The finding of the surprisingly small mass of Sagittarius A* extends to the problems beyond our galaxy as well.

Resolving Dark Energy

Dark energy was established to explain why galaxies were not crashing into each other. That problem no longer exists. Based on the relatively small masses which black holes like Sagittarius A* contain, the extent of a black hole's gravitational influence is quite limited.

Dark energy is not needed as an anti-gravitational force between galaxies *because 'there isn't any gravity there.'*

Paradox Null
~~Galaxy Gravity~~
~~Dark Energy~~

How was the Mass of Sagittarius A* Calculated?

The mass of Sagittarius A* was calculated based on the Earth's annual temperature variations.

That may surprise you.

The solution to resolving the mass of Sagittarius A* was revealed while solving the paradoxes of Earth's temperature.

There are two broad paradoxes related to Earth's seasonal and persistent surface temperatures – the *Cold Perihelion Paradox* and subterranean thermal paradoxes.

THERMAL PARADOXES

The Earth receives, based on popular estimates, 95% of its thermal energy from the Sun. The Earth's exposure to solar radiation increases when the Earth gets closer to the Sun. The Earth receives 3% more solar radiation when close to the Sun – at perihelion – than when farthest away from the Sun – at aphelion.

The Cold Perihelion Paradox

Based on the Earth receiving 95% of its thermal energy from the Sun, the Earth should be a lot warmer when the Earth is closer to the Sun. The Earth normally (at its semi-major axis) receives 2.46E+17 watts of power from the Sun. If the Earth generates 5% of its own heat, it generates 1.23E+16 watts. The average surface temperature of the Earth is 288° Kelvin. Add the input wattage up and divide it by the temperature to see how much heat it takes to stay warm.

$$(2.46E+17 \text{ watts} + 1.23E+16 \text{ watts}) / 288° \text{ Kelvin} = 8.55E+14 \text{ watts per Kelvin}$$

$$2.58E+17 \text{ watts} / 288° \text{ Kelvin} = 8.55E+14 \text{ watts per Kelvin}$$

Based on this model, the Earth's temperature rises or falls 1 Kelvin per each unit of 8.55E+14 watts of power. Now let's calculate Earth's temperature with the wattage Earth receives at perihelion and aphelion.

2.66E+17 watts / 8.97E+14 watts per Kelvin = 296.53 Kelvin

2.50E+17 watts / 8.97E+14 watts per Kelvin = 278.22 Kelvin

The total difference in temperature expected on Earth at perihelion and aphelion is 18.3 ° Kelvin.

296.53 ° Kelvin – 278.22 ° Kelvin = 18.31 ° Kelvin

If Earth receives 95% of its heat from the Sun, Earth should be 18.3 ° Kelvin – 33 ° Fahrenheit – warmer when closer to the Sun. That should be the whole planet average.

But it's not.

In fact, the Earth is 2.3 ° Kelvin <u>colder</u> at perihelion than at aphelion. That is a variance from expectation of 20.6 ° Kelvin, 35 ° Fahrenheit.

Paradox simple: If the Sun provides 95% of Earth's heat, than Earth is not 2.30° degrees colder at perihelion, but it is.

It is impossible that the Sun provides 95% of Earth's heat.

It's Hotter Underground Paradox

If the Sun provides Earth with 95% of Earth's heat, then the surface of the Earth must be the hottest place on Earth.

It's not.

Some researchers suggest that the hot interior parts of the Earth are due to latent heat – heat remaining from the original big bang. That theory fizzled out when big bang theory was pushed back past 2 billion years ago. Lord Kelvin calculated that that heat would have run down to present-day levels after merely 100 million years.

The big bang heat idea was replaced with heat-left-over-from-pre-Solar-System-supernova theory. That failed because the Earth kept insisting that it was more than 4 billion years old. That made the same problem for the supernova heat theory as the big bang heat theory – there just wouldn't be any heat left.

So then the Sun got involved and, using stellar fusion theory, it burned super-super hot before cooling down to present-day levels.

But there were volcanoes on Earth's surface 4.4 billion year ago. The famous 4.4 billion-year-old zircon crystal was formed in a volcano.

That means that the surface of the Earth was thoroughly solid at that point in time, so the 'hot Sun' idea – if it had any basis at all – wouldn't really explain anything. That 'super-heat' would have had to have happened over 4.7 billion years ago.

And what about the Sun? Via the Kelvin-Helmholtz mechanism the Sun got started about 8.9 million years ago.

So if Earth formed in a molten state, what's been powering volcanoes for the last 4.6 billion years after the all the Kelvin-Helmholtz heat was squeezed out to 'present-day levels'?

The latent heat ideas were later amended with heat from radioactive metals.

Radio-Metals Paradox

The radio-metals idea is that the radioactive metals uranium, thorium, and potassium convert themselves from mass into energy, collectively generating, 4.4E+13 watts of heat inside of the Earth.

That's about 0.0184% as much heat as the Earth receives from the Sun. That's not nearly enough heat. That theory probably emerged before deep mines and deep bore holes existed, or the researcher did not consider such information.

You see, if you dig a hole in the summer-time to 500 meters deep, everything gets colder and colder the deeper you go. The depth of 500 meters is the maximum depth at which the radio-metals-heating-the-Earth theory works. If Earth is generating merely 0.0184% of the heat received from the Sun, things should continue to get colder as you go deeper.

Instead, things start getting warmer. From ~500 meters on down, temperatures rise 25-30 Kelvin for every kilometer descended. The highest non-volcanic temperatures ever directly observed were 650° Kelvin or 710° Fahrenheit.

Earth's highest temperatures are observed in volcanic activity and in 'hot spots' beneath volcanoes. The physical temperature limit of a volcano or hot spot occurs between 1,900° Kelvin and 2,300° Kelvin. Those are the melting temperatures of wet granite and dry granite respectively. When temperatures exceed ~2,000° Kelvin, the granite cap of the hot spot melts thin and the volcano erupts.

The theory of heavy metals inside of Earth presently generating heat is sound. But that does not provide nearly enough energy for the observed thermal effects, nor nearly enough heat to explain the very regular temperatures on Earth's surface.

It's a Lot Hotter Down Below Paradoxes

Some theories hold that the center of the Earth is over 7,000° Kelvin.

That is not based on the real, actual, observed thermal gradient of 25°-30° Kelvin. That is based on a *theoretical* geothermal gradient ten times that high, beginning at depths never physically achieved.

To put 7,000° Kelvin in perspective, here are the boiling points of some chemicals at Earth-surface atmospheric pressure.

Super-hot core theories have a lot of problems. Among them:

The Curie Paradox

If the core is hotter than 1,043° Kelvin, then that region is too hot to maintain any magnetic field. Earth certainly does have a magnetic field. Iron and all other ferrous elements do not maintain any magnetic field when hot, molten, or gaseous.

The Light Elements Paradoxes

If the core is hotter than 3,000° Kelvin, there cannot be any light elements – oxygen, helium, hydrogen, or nitrogen – down there.

Since those should have all been refined away, we should find purified metals beneath Earth's surface – refined iron, aluminum, silicon, and calcium metals. That never happens.

The Liquid Paradox

Matter under extreme pressure must exist in its densest state. If matter cannot exist in its densest state due to high temperatures, then fission occurs.

To get a concept of what to expect subterranean temperatures to look like, look at Earth's lakes and oceans.

Deep down in deep lakes and ocean canyons, water, when experiencing the most extreme pressures documented, remains a liquid. Water remains a liquid when under extreme pressure because water is <u>less</u> dense as a solid than as a liquid. Fresh water at the bottoms of deep lakes remains a liquid and never freezes.

That water, and also deep ocean water, also remains below the freezing point of water. That densest state of water is cold and liquid.

Virtually every other chemical has a solid densest state. Deep Earth must exist in its densest state: cold and solid.

At least solid.

And if it's solid, then the maximum temperature must be lower than the melting point of all its major components, including silica.

Silica, as the main ingredient in granite, has a melting temperature of 1,900° Kelvin to 2,300° Kelvin. That, then, is the absolute maximum temperature at which subterranean matter can exist. Actual highest temperatures are probably half that, around 1,000° Kelvin or 1,340° Fahrenheit.

Based on the Russian Super-Deep Bore Hole trend, that maximum temperature would occur at a depth of approximately 36 kilometers.

And that's a problem.

Thermal Limitations

If the average temperature of Earth below 36 kilometers is merely 1,340° Kelvin, then Earth's total thermal capacity is merely 1/5 of the capacity formerly estimated.

Then there's also an iron core problem. There is no direct evidence that any iron core exists. Actual drilling and sampling found that Earth is made of the same stuff all the way down – water, granite, sand, CO_2, methane, bacteria, and algae down as far down as we could get. Crude oil, too.

Based on direct evidence instead of inferred theory, the Earth's interior is granite instead of iron and nickel. Granite is far lighter and has a far lower thermal capacity than iron or nickel.

Now the total thermal capacity of Earth is reduced to 1/10 of former estimates. Now big bang heat and supernova heat would run out in less than 80 million years. Now, the surface of the Earth would be the hottest place on Earth.

2.58E+17 / 7.58E+16 watts per Kelvin = 3.4 Kelvin

The surface of the Earth would be about 3.4° Kelvin (-453° Fahrenheit).

This leads to two conclusions; 1: The Earth continually generates vast quantities of heat, and 2: That heat is more than nuclear decay.

Why so warm?

THE LAW OF ATOMIC ENERGY TRANSFORMATION

Atomic energy transformation (AET) occurs when proto-electromagnetic radiation (zero-point energy/ultra-high frequency gamma radiation) affects an atom, is transformed by that atom, and that energy is then re-emitted. This is transformation of electromagnetic and/or proto-electromagnetic energy.

We can predict how much energy an object will emit through Atomic Energy Transformation with the following equation.

$$\text{Mass} * \text{Time} * H = \text{Wattage}$$

H is the James Harris Energy constant.

$$H = 2.489E\text{-}030$$

The energy transformed through AET is, in large part, energy emitted from black holes. Black holes are very high-frequency objects compared to Earth, and their emissions are higher frequency (shorter wavelength) than light.

When wavelengths get very short – less than twice the width of an atomic nucleus – those wavelengths are no longer electromagnetic. They are no longer electromagnetic at that width because when a wavelength is too short to affect a whole atom, that wave <u>will not induce any photon or elec-</u>

tron. There is not, therefore, any *electromagnetic reaction*.

But very short waves *do* affect the individual parts of atoms. The transformations and inductions of these energies flip quarks. When the subatomic charges of the quarks flip (change 'color' and 'flavor'), that shifts the nuclear (atomic) magnetic field of the atom. The shifting of that magnetic field, in turn, induces and destroys electrons in the electron shells.

This is the experience of time. This is a working *definition* of time.

In a cesium atom, those inductions and destructions occur 9,192,631,770 times per second.

When electrons are destroyed, they emit electromagnetic radiation – light waves. Cesium atoms destroy electrons 9,192,631,770 times per second, so cesium emits radiation at a constant frequency of 9,192,631,770 Hz.

That is the international definition of one second.

Atomic Energy Transformation happens at the speed of time. Atomic Energy Transformation happens because of time.

SOLVING EARTH'S TEMPERATURE PARADOXES: AET ON EARTH

The Earth is 650 kilometers thick based on seismic reflection data and the angle of gravity. With that, the mass of Earth has been calculated as 9.56E+23 kilograms. From this mass, we can calculate the AET of Earth.

$$\text{Mass of } 9.56E+23 \text{ kg } * \text{ Time } 9.07E+24 \text{ kg } * \text{ H } 2.489E-30 = 2.16E+19 \text{ watts}$$

According to the Law of Atomic Energy Transformation, Earth generates 2.16E+19 watts of power.

That is 80 times more energy than Earth receives from the Sun.

Energy emitted from radioactive elements is energy generated through atomic energy transformation. Those emissions are included in this number.

All of this energy ultimately leaves through the surface of Earth. Therefore, we can add up the total energetic input of Earth, and that will be equal to the total output of Earth.

$$\text{AET power } 2.16E+19 + \text{Solar input at semi-major axis } 2.46E+17 = 2.18E+19 \text{ watts}$$

Star contributions to Earth are also counted. Star contributions are negligible, totaling 1.16 billion watts or 0.000000005% of Earth's total input.

To find out how output wattage translates to surface temperature, divide the total output of Earth by the average surface temperature of Earth to determine how many watts equal how many Kelvins of heat on the surface of Earth.

$$2.18E+19 \text{ watts } / 288° \text{ Kelvin } = 7.58E+16 \text{ watts per Kelvin}$$

For every unit of 6.93E+16 watts that Earth outputs, the temperature rises 1° Kelvin on the surface.

PROVE IT

The claim that Earth generates 2.16E+19 watts of power contradicts popular theory. Within this model, Earth receives merely 1.125% of its energy from the Sun and generates 98.88% of the energy it experiences.

Because Earth generates most of its own heat, the difference between the total energy Earth experiences at perihelion versus aphelion is rather small.

> AET power 2.16E+19 + Solar input at perihelion 2.54E+17 = 2.1841E+19 watts

> AET power 2.16E+19 + Solar input at aphelion 2.37E+17 = 2.1824E+19 watts

Divide these numbers by Earth's watts-per-Kelvin constant to predict Earth's average temperature.

> 2.1841E+19 watts / 7.58E+16 watts per Kelvin = 288.10° Kelvin

> 2.1824E+19 watts / 7.58E+16 watts per Kelvin = 287.88° Kelvin

Instead of an expected orbital variation of surface temperature of 18.3° Kelvin, the expected variation is much smaller.

> 288.11° Kelvin - 287.87° Kelvin = 0.22° Kelvin

If Earth generates 98.88% of its own heat, than the expected orbital temperature variation is merely 0.22° degrees Kelvin.

But that does not entirely solve the riddle. Earth is not 0.22° Kelvin warmer at perihelion. It is 2.30° Kelvin colder.

The Law of AET is not the only new law at play here. To finish resolving this riddle, calculate the Law of Time.

The Law of Time is that the speed of time is proportional to mass proximity. This matters to Earth's temperature because Earth's orbit is not circular. When Earth gets closer to the Sun – culminating at perihelion – the Earth experiences higher mass proximity than at aphelion. Therefore, time happens faster when Earth is at perihelion than when Earth is at aphelion.

How much faster? 0.871% faster.

What does that mean to temperature?

That means nothing to Earth's AET or self-generated power. AET happens at the speed of time. Earth's time speed could occur in slow-motion or fast forward and Earth's experience of its own heat would remain constant.

That means that Earth's orbital changes in temperature are experienced due to the Sun.

So, how can temperatures go down when we are closer to the Sun?

That happens because the temperature of the Sun goes down. Since the speed of Earth-time increases when Earth approaches the Sun, the *apparent* temperature of the Sun falls. The peak output frequency falls. The Sun get colder *relative to Earth.*

We can calculate how much colder the Sun is at perihelion than at aphelion. To accomplish this, divide the temperature of the Sun (constant) by the specific TCF of Earth – that is, divide the Sun's temperature by the TCF of Earth at various points in Earth's orbit.

5,778° Kelvin / semi-major axis TCF 1.000 = 5,778° Kelvin

5,778° Kelvin / perihelion TCF 1.004 = 5,755° Kelvin

5,778° Kelvin / aphelion TCF 0.995 = 5,806° Kelvin

The Sun is 0.877% colder – *as perceived and experienced by Earth* – when Earth is closest to the Sun – when Earth is at perihelion. That's 51° Kelvin or 91.8° Fahrenheit colder.

That affects the <u>actual wattage</u> the Earth receives from the Sun.

AET power 2.16E+19 + (Solar input at perihelion 2.54E +17 / TCF of 1.004) = 2.1754E+19 watts

AET power 2.16E+19 + (Solar input at aphelion 2.37E +17 / TCF of 0.995) = 2.1929E+19 watts

Then we can predict the average surface temperature of Earth as locally experienced. Divide the total wattage by Earth's watts-per-Kelvin constant.

2.1754E+19 watts at perihelion / 7.58E+16 watts per Kelvin = 286.96° Kelvin

2.1929E+19 watts at aphelion / 7.58E+16 watts per Kelvin = 289.26° Kelvin

Based on the Law of Atomic Energy Transformation and the Law of Time, the expected orbital temperature variation between perihelion and aphelion on planet Earth is -2.30° Kelvin.

286.95° Kelvin perihelion - 289.25° Kelvin aphelion = - 2.30° Kelvin variance

And that is exactly what observations report.

Atomic Energy Transformation

Paradox Null
~~Radio-Metals and Latent Heat Paradoxes~~
~~It's Hotter Underground Paradox~~
~~It's a Lot Hotter Down Below Paradox~~
~~The Curie Paradox~~
~~The Light Elements Paradoxes~~
~~The Liquid Paradox~~

The Law of Atomic Energy Transformation is a very robust theory

Can it handle the biggest, most wondrous, most awe-inspiring object known to mankind? Can it handle the Sun?

NEW SUN RISING

The Sun is popularly taught as running on fusion. We have known since just two years after that theory was proposed that the surface of the Sun is far too cold for that – merely 5,778° Kelvin or 0.039% of the 15 million degrees needed to (theoretically) maintain fusion reactions.

The author's concept of the Law of Time exacerbates this problem. The Sun has a TCF of 9.39. The surface temperature of the Sun *as locally experienced* is 619° Kelvin/655° Fahrenheit. The surface of the Sun is colder than the surface of Venus.

There is also the *It's a Lot Hotter Down Below Paradox.* That applies to the Sun, the Earth, and every other object greater than about 100 kilometers in radius. Subterranean temperatures can only rise so much under such pressures – up to about 1,000° Kelvin.

If temperatures got nearly as hot as popular models predict, and if stars were made of hydrogen, then stars would just boil away.

What if all of the Sun's power was the result of Atomic Energy Transformation?

Sun mass 1.99E+30 kg * Sun time 9.25E+25 * Ħ 2.489E-30
= 3.84E+26 watts

That is the Sun's total output.

Using the <u>exact same equation</u> which is proven correct for Earth and its temperatures, the Sun's output is calculated to within 0.16% of the popularly published output of 3.85E+26 watts.

The Sun is powered through Atomic Energy Transformation.

Paradox Null
~~Stellar Fusion Paradoxes~~
~~Stellar Suicide/Supernova Paradoxes~~
~~Stellar Evaporation Paradoxes~~
~~Boiling Star Paradoxes~~
~~Fission Paradoxes~~

AET, SIMPLE OVERVIEW

The power the Sun emits is generated through the mass of the Sun experiencing time. That energy is then re-emitted into the universe where, at some point, something else absorbs that energy and re-emits that energy. The difference between AET and ordinary transformation of energy is indistinct.

AET is characterized by the unusual effects of
1. causing atoms to have electron cycles
2. causing atoms to generate heat
3. super-heating the coronas and upper atmospheres of stars, planets, and moons,
4. heating the subterranean regions of stars, planets, and moons, and,
5. preventing absolute zero-degree conditions.

The source of atomic energy transformation is: everything.

The recipient of atomic energy transformation is: everything.

The differences between electromagnetic and proto-electromagnetic radiations are the frequency at which they occur, the depths to which they penetrate, and the scales of their various interactions.

MASS AND ATTRIBUTES OF SAGITTARIUS A*

The mass of Sagittarius A* was revealed through calculating how the Earth is 2.30 degrees Kelvin cooler at perihelion than at aphelion. If Sag A* was not factored into the equation, the temperature on Earth was predicted to be 3.84° Kelvin cooler. If Sag A* was factored in at 4.31 million-solar-masses, the temperature predicted was just 0.01° Kelvin cooler.

The mass of Sagittarius A* was thereby evidenced through the effects of Sagittarius A* on speed of time on Earth, as evidenced through observed orbit-related temperature fluctuations on the surface of Earth.

Factors considered within this model included the albedo of Earth and variability of albedo. Research indicated no significant change in Earth's net albedo based on either seasons or orbital period.

This method of determining the mass of Sagittarius A* is far too sensitive to allow for significant deviation. This model of predicting the mass of Sagittarius A* is believed to be accurate to within 10% - Sagittarius A* has a total mass of between 29,000 and 24,000 solar masses.

This eliminates the theoretical needs for both dark matter and dark energy.

Paradox Null
~~Galactic Gravity Paradoxes~~
~~Intergalactic Gravity Paradoxes~~
~~Dark Matter~~
~~Dark Energy~~

As with other predictions of *Paradox Null*, these predictions are testable in any number of ways. The peak emission frequency of Sagittarius A*, for instance, can be predicted based on the speed of time there.

Big, Bright Sagittarius A Star

Using a fairly simple model, the peak emission frequency of Sagittarius A* can be estimated based on the Sun's peak emission frequency.

Assuming that the peak emission frequency of Sagittarius A* is proportional to the speed of time, and assuming that the structure of Sagittarius A* is similar to the structure of the Sun, multiply the Sun's peak emission frequency by the Sagittarius A*'s SU (Solar Units of time) value of 4,265.11 to predict Sag A*'s peak emission frequency.

$$6.00E+14 \text{ Hz} * 4,087.1 \text{ SU} = 2.45E+18 \text{ Hz}$$

This model predicts the peak emission frequency of Sagittarius A* to be 2.45E+18 Hertz. That frequency occurs in the upper x-ray range. Sagittarius A* is expected to emit radiation primarily in the x-ray spectrum.

Based on the NASA Chandra photograph of Sagittarius A*, it appears that Sagittarius A* does, indeed, emit x-rays.

Based on the output frequency predicted, the surface temperature of Sagittarius A* as observed from Earth is 24,700,000° Kelvin or 44.5° million Fahrenheit. That same temperature, *as locally experienced* is much lower.

24,700,000 ° Kelvin / TCF of
38,133 = 647.7 Kelvin or
706.2 ° Fahrenheit

That makes sense.

That makes sense because Sagittarius A* should be colder than Curie temperatures to have much of a magnetic field. Sagittarius A* arguably has a powerful magnetic field.

That makes sense because if it was thousands of degrees hotter, Sag A* would boil away.

And if Sagittarius A* is an ordinary, solid body, then it does not suck in atmosphere. Instead, it should maintain a vast cloud and emit large amounts of gases.

As shown in this picture by NASA Chandra, Sagittarius A* does indeed maintain a large cloud of gases.

This cloud is not disc-like. This cloud is three dimensional and averages as roughly circular. The billowing forms of the outer reaches of the cloud indicate that this cloud is outbound. Gas density rises near a disc of debris near the center. Since this cloud is not flat, this cloud cannot be orbiting Sagittarius A*. Instead of orbiting or flowing into Sagittarius A*, this cloud is flowing outwards.

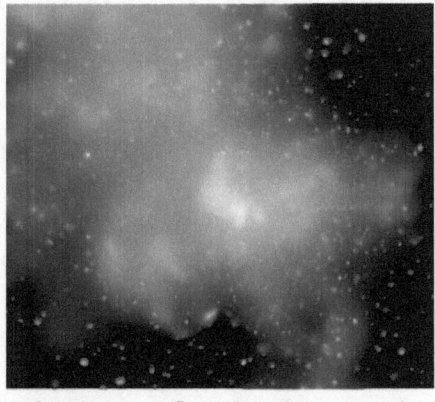

If that black hole in the middle of this picture was the medieval, magical black hole of old, that bright spot in the middle should be dark and cold and all those non-orbital gases should be have been sucked in already.

Sagittarius A* has been detected as emitting temperatures of 100° million Kelvin when collisions have occurred.
How hot was that? To find the temperatures as locally experienced, divide the apparent temperature by the TCF.

100,000,000° Kelvin / TCF of 38,133 = 2,622° Kelvin

A significant collision with Sagittarius A* generated temperatures of 2,622° Kelvin – temperatures very similar to the temperatures generated by a volcanic eruption. Those are temperatures consistent with an ordinary solid planet impacted by an ordinary solid planet.

Those temperatures are about 4 times the temperature of Sagittarius A* normally. This, too, makes perfect sense. Applying the same model to Earth, Earth's average surface temperatures would rise to 4x current levels if a proportionally similar impact occurred.

Is it possible that if the Moon made direct impact onto the Earth that surface temperatures would rise from an average of 288° Kelvin (59°F) to 1,152° Kelvin (1,614°F)? Yes. That is possible.

Would temperatures get higher than that? At about that temperature, sodium boils and water, potassium, and sulfur are boiled or boiling – all of those boiling actions remove heat from the planet and increase the gaseous discharges from it. Those evaporations limit the temperature. That quasi-maximum temperature makes perfect sense.

Hawking Hole Temperature Paradoxes

If, however, Sagittarius A* were 4.31 million solar masses, the TCF would be 142,616,049 and the temperature of 100° million Kelvin <u>as locally experienced</u> would be just 0.7 Kelvin.

Temperatures that low are very unrealistic. There would not be any explosion. Even hydrogen and helium would not evaporate, expand, or explode at that temperature.

And if Sagittarius A* were 4.31 million solar masses with Einsteinian time dilation – where time comes to a screeching halt at the corona aka 'event horizon' – the region observed might

be occurring at merely 1/10,000th the speed of Earth. *That* would mean that those temperatures perceived as 100° million would be – as locally experienced – 10,000 times higher or 10° trillion Kelvin.

Both of those temperatures – 100° million and 10° trillion – are entirely paradoxical because of themselves. Hydrogen and helium fission at lower temperatures than that. When hydrogen and helium fission, they are endothermic. Hydrogen and helium fission events lower temperatures.

Before temperatures get to 1 trillion, every other element also fissions.

Of course, if anything got nearly that hot, it would all evaporate away before anything else happened.

If anything got nearly that hot, there wouldn't be anything left to *be* hot.

All temperature observations related to black holes contradict Hawking Hole theory.

Paradox Null
~~Event Horizons~~
~~Temperature Paradoxes~~
~~100,000,000° Kelvin explosion~~
~~1° trillion Kelvin explosion~~
~~0.7° Kelvin explosion~~
~~Fission Paradoxes~~

Hawking Hole Gravity Paradox

There is also a huge problem when it comes to gravity. Gravity happens at the speed of time. One Earth G = 9.8 Newtons _per second._ If Sagittarius A* experienced no time, then the mass of Sagittarius A* would exert NO GRAVITY.

The surface of Sagittarius A* generates 6,212 Earth Gs of gravity.

 6,212 Earth Gs per second * 0 (zero) seconds = 0 (zero) gravity

Since black holes do exert gravity, it is impossible that time does not occur there.

Event horizon theory is a myth.

Black holes are really just ordinary objects which experience time very quickly and emit energy at very high frequencies.

Paradox Null
~~Time Paradoxes~~
~~Gravity Paradoxes~~

Black Hole Power

Now let's figure out how bright Sag A* is with the Law of Atomic Energy Transformation.

$$\text{Mass of Sag A* } 4.61E+34 \text{ kg * Time } 3.46E+29 \text{ kg *}$$
$$Ħ \, 2.489E-30 = 3.97E+34 \text{ watts}$$

Sag A* puts out 104 million times more power than the Sun does.

Black holes do not suck in and destroy energy. Black holes transform energy just like everything else does.

Paradox Null

Information Paradox

Accretion Disks

According to classic black hole theory, the accretion disk is the black hole buffet of stuff that gets sucked in when the black hole is hungry. Accretion disks are thought of as being different than stellar disks. When an accretion disk looks hot, it is thought of as experiencing Kevlin-Helmholtz compression. When a stellar disk looks hot, it is thought of as reflecting and re-radiating heat from the star in the middle.

This indicates that the interior of a Hawking hole would be cold. Absolutely cold. Zero degrees cold.

By extension, that same logic indicates that the interior of the Sun would be cold. Very cold. And frozen after 8.9 million years.

And what is there to compress anyhow? Rock are rocks. Rock do not get significantly more dense under pressure. Only the gases can be compressed. Gases are low-density, so they don't contain much heat and don't have much to contribute.

Since gases are observed flowing out of Sagittarius A*, however, those gases are not being compressed and the Kelvin-Helmholtz mechanism has no role in this situation. Accretion disks generate power through atomic energy transformation just like everything else does.

Does Heat Repel from Gravity?

The idea that heat is repelled from gravity is basically true. It happens by heat being convected out and away from solid objects, especially through evaporation. The hotter the gases, the lighter they are, so heat is moved away from centers of gravity through evaporation and convection of gases. Heat is also radiated away.

The hotter an object is, the faster all of these things happen.

The fact that Earth and other objects still have heat inside after at least five billion years evidences that every object generates heat through atomic energy transformation.

ATOMIC ENERGY TRANSFORMATION AND THE LAW OF TIME

As seen in the numerous examples above, the Laws of Time and AET are maxims. They are true because of themselves. There are anti-paradoxical. They are the opposites of paradoxes. They are the cures to paradoxes.

Like Newton's laws, which opened new frontiers into physics and the cosmos, these two laws make the universe comprehensible, definable, and explorable in ways not formerly imagined.

With those two equations, the mass of Sagittarius A* is found to be relatively light. That solves the mysteries and paradoxes of both dark matter and dark energy.

Finding Sagittarius A*'s peak emission frequency to be 2.56E +18 Hz is consistent with observations, and solves a series of the mysteries concerning black holes: There is no event horizon. There is no singularity. There is no inescapable gravity. There are no flying photons. There is no General Relativity.

REWRITING
STANDARDS

When describing describing cosmic distances, distances should be described in terms of parsecs. "The universe as estimated to a distance of 10 billion parsecs," for example. A parsec is a physical, triangulated distance equal to $9.46E+12$ physical kilometers.

It is important to distinguish 'physical meters' from 'meters'. The standard definition of a meter, starting in 1987, is 'the distance light travels in $1/299{,}792{,}458^{th}$ of one second. That is not a definition of distance. That is a definition of the speed of light. The speed of light is variable. This definition of the meter is inherently misleading.

Likewise, the use of the term 'light-year' is erroneous. The speed of time in intergalactic space falls as low as $1/50^{th}$ the speed of Earth-time. A 'distance of one light year' in that space, then, is much shorter than formerly imagined. The speed of light is much slower there.

$$299{,}792{,}458 \text{ meters per second} / 50 =$$
$$5{,}995{,}849 \text{ meters per second}$$

Intergalactic light, as perceived from Earth, travels as slow as 5,995,849 meters per second.

So during one Earth year, intergalactic light travels as little as $1/50^{th}$ of one 'light year'.

Light is not a cosmic measuring tape, it's a bungee cord. The term 'light year' is misleading and obsolete. Since one light year implies that *the light of that transmission* is of a certain age, the term 'light-year' is entirely misleading.

The *average* speed of intergalactic light is roughly 1/25th the speed of Earth-speed light.

Ages of objects outside of the Milky Way and Andromeda galaxies may be estimated by multiplying their distance in light years by 25 and translating the assumed light-years into parsecs.

Example: If a report states that an event happened 1 billion years ago and 1 billion light years away, then that event occurred approximately 25 billion years ago 0.3 billion parsecs away.

It will take years to write that term out of academia. Start now.

3.26 light years at Earth-speed time = 1 parsec

NEW HORIZONS

Hubble's Law should not be used for... anything.

Fitzgerald-Lorentz compression does not exist.

According to the Law of Atomic Energy Transformation, energy <u>output</u> is proportional to mass. Energy does not equal mass proportional to the speed of light.

The proto-electromagnetic energies which drive AET may be harnessed through electromagnetic arrangements and transformed into useful power. A conceptually simple version of this is to build 'solar panels' which are tuned to very high frequencies. I have formerly referred to very high frequency energies as zeta-rays so call the panels zeta-panels. Zeta-panels can generate power any time of day or night, may be installed anywhere. They would work for power generation in mines, and would serve as a superb power source for satellites and explorers.

Gravity, as a proto-electromagnetic force, may be electromagnetically manipulated, induced, and reduced. Space travel and space exploration will never be effective until this reality is embraced.

Powerful, newer telescopes can accurately determine distance through focus triangulation. Galaxy maps should be drawn and redrawn based primarily on information obtained through triangulation, parallax, and Harris Blue-Shift.

Masses of distant galaxies can be effectively estimated using the Law of Time and Harris Blue-Shift. The color of a galaxy indicates the mass proximity of that galaxy. Since a typical galaxy has proximity only to itself, mass proximity indicates total mass in most cases. The mass of the galaxy indicates the size of that galaxy. Also factor in the spread of the galaxy or 'galactic density'. If a galaxy is spread thin, than it has more mass than a denser galaxy of the same color. If a galaxy is very dense, it has less mass than looser galaxy of the same color.

This way, the size of a galaxy can be determined primarily from its color. The size of the galaxy can then be compared to the visual size of the galaxy to calculate distance.

Every planet is realized to have a solid surface including Jupiter, Saturn, Sol, and Sagittarius A*. Explore them.

Star masses, especially those stars near galaxy centers, can be recalculated to include consideration of the Law of Time. Since the inner galaxy experiences time more quickly, the blue stars of the inner galaxy may be merely 10% of masses formerly estimated. Black holes may be merely 0.54% of former estimates.

Supernovae are collisions. Collisions come in many shapes and sizes – so many, in fact, that no two are alike. Treat supernovae-based reports with skepticism.

Gravity waves result from events of large-scale fission and cause intermittent dilation and contraction of local time locally.

Matters of planetary structure, geology, evolution, and biology are generally beyond the scope of this book. The author's basic conclusions are that the Earth has little mass inside, that Earth's radius continues to increase in consequence of Earth's binary orbit with the Moon, that all mountains are volcanic in origin, that Earth's evolving climate dictates changing biodiversity, and that virtually every object in the universe support bioactivity. Watch for future books.

CONCLUSION

Paradox Null greatly simplifies physics.

The Speed of Light

The speed of light is variable. This eliminates Lorentz compression, the twins paradox, the train paradox, time-travel paradoxes, high-speed travel paradoxes, and lensing paradoxes.

Lensing

All lensing consists of the changing of the speed of light.
Lensing involves two phenomenon: The changing of light speed (Harris Blue-Shift) and wave-particle-wave transformation delay.

Lensing and gravitational lensing happen for the same reason.

Inertial Frame of Reference

The inertial frame of reference consists of discreet subatomic particles - aether-pi - in a gaseous state. Actions of this field with respect to matter include time, inertia, gravity, light, and magnetism.

Actions of this field with respect to atomic physics include electron action, photon action, electromagnetic action, atomic energy transformation (AET), wave-particle transformation (WPT) the weak force, the strong force, the electromagnetic force, time, quark flipping, quark coloration, gravity, gluon formation, and Heisenberg uncertainty.

The speed of light is constant within a macro-atomic inertial frame of reference. This is not true on atomic or planetary scales, where the speed of light varies proportional to distance.

Black holes, Stars and other Objects

All objects experience time proportional to their mass proximity. Black holes and stars are ordinary planets in that they have cohesive, solid surfaces. They are dissimilar to small planets and moons in that they have very deep atmospheres.

All spheres form from solid debris. All spheres exist as solid bodies. Virtually all spheres form with rotation.

All objects emit energy proportional to their mass and experience of time. Black holes emit very high frequency emissions. Star emit lower frequencies. Planets emit still lower frequencies. Every object obeys the same laws. There is no stellar fission or fusion.

The masses of black holes are significantly lower than formerly estimated and their gravity is easily escaped.

Quasars are black holes with heavy debris fields around them where collisions frequently occur or did occur.

Pulsars and Polar Jets

All spheres exhibit polar jets. The polar jets of Jupiter and Saturn erupt from openings so large the Earth can fit inside. Famously both Jupiter and Saturn each have a visibly hexagonal polar opening – an evenly six-sided hole.

If a sphere is expanded and not dense, the poles of the sphere are weakly supported, loosely packed, and frequently present openings to the planet's interior. Polar jets erupt from the natural poles.

A pulsar exists when a sphere's rotation has been disturbed such that the polar jets no longer exist on the rotational axis. When that happens, the polar jet shines like a beacon on top of a light-house, flashing at twice the speed of the pulsar's rotation. Each rotation shows the north pole once and then the south pole once.

Once again, all celestial bodies obey the same rules.

Neutron Stars

Neutron stars consist of the cores of stars or black holes. The core of a planet, star, or black hole typically consists of the following.

Palladium 41%
Platinum 14%
Gold 11%
Lead 9%
Thorium 6%
Osmium 4%
Iridium 3%
and other trace elements.

During planetary formation, these elements become the core of planet which is orbited. The planet orbits its core. The planet at large is never molten, but the core does become molten and, due to very high temperatures, refined. The core weighs five times more per cubic centimeter than the bulk of the planet does. The palladium-based core weighs 15.19

grams per cubic centimeter.

The planet may be stripped away from it core, as through a major collision, leaving a 'neutron star', or core behind.

The concept of 'neutron stars' such as are *composed of neutrons* is fictional. These object consist primary of palladium and are properly called *palladium stars* or *core stars*.

The Universe

Movement

Cosmic expansion is not observed as existing. Spectral shifts are most accurately comprehended using Harris Blue-Shift.

The imagined necessity of intergalactic movement arose first from mythology, then from a false equation founded on false premises, then from misinterpretation of extra-galactic spectroscopy, and then from the overestimation of the masses of black holes whereby galaxies were believed to have significant gravitational influence on each other.

Since time is proportional to frequency and size is proportional to time, the more massive a galaxy, star, or any object, the higher its emission frequency. Therefore little galaxies are red. Red galaxies are <u>not</u> big, white galaxies far far away flying quickly away. They are just small.

Whereas black holes weigh merely 0.05% of what has been formerly estimated, the gravitational interactions between galaxies are not significant. Since galaxies have little gravitational effect on each other, galaxies need not be moving to avoid crashing.

Age

The universe has existed as presently observed for not less than 1.25 trillion years as indicated by the Hubble Telescope's Extreme Deep Field view of galaxies 13 billion parsecs away given that the speed of intergalactic light averages $1/25^{th}$ the speed of light on Earth.

Given the findings of Fritz Zwicky, the age of the Milky Way galaxy is not less than 2E+19 years.

The universe is probably older than the Milky Way galaxy as defined above.

The is no real or practical evidence indicating that the universe began, that the universe has any definite age, or that the universe will ever end.

CMB

The cosmic microwave background originates with galaxies. There are so many galaxies that every point in the sky may contain a galaxy at some distance.

Extent

The extent of the universe is unknown. The universe is apparently limitless.

Accretion Disks

Accretion disks, stellar disks, planetary disks, lunar disks, and asteroid belts are all the same phenomenon of debris caught in orbit. If the disk is dense enough, spheres form from it.

Nucleogenesis

Nucleogenesis occurs on, in, or about every significant cosmic object and is evidenced by the continual emission of 'solar winds' from every known planetary, stellar, and black hole body. Structures most likely to induce nucleogenesis (via wave-particle transformation WPT) include radioactive elements, photosynthetic chemicals, and DNA.

Life

All spheres are solid and formed solid. Debris forms when spheres collide. The more violent the collision, the more widely debris spreads. Those debris eventually form or contribute to other spheres. At no point in this creation/destruction cycle has the debris been purged of water bears, bacteria, anaerobic bacteria, algae, or any simple life form.

Debris which join a planet early in formation do not fall through any atmosphere, do not experience any significant acceleration, and typically do not experience significant impact.

Life evidences itself via the emission of hydrogen and helium and through the presence of atmospheric oxygen, nitrogen, and /or volatile organic compounds. Atmospheric oxygen, nitrogen, and volatile organic compounds (VOCs) are produced exclusively through biological activity. These chemicals variously exist on every known planet, every measured moon, and the Sun.

Evolution

As with the age of universe, the beginning of life is unknown. As it is observed that life is the most likely source of nucleogenesis, it would seem that life began the universe, and not the other way around.

As for planet Earth, the radius of Earth is increasing. The mass of Earth is not increasing. As the Earth spreads, the atmosphere of the Earth grows thinner and thinner. Because of this, plant and animal life on Earth is decreasing in both size and frequency. Earth's climates grow colder and colder. Localized, recurring extinctions of dinosaurs occurred primarily in consequence of mega-eruptions and meteor-falls, but the complete extinction of the dinosaurs occurred when Earth's atmospheric pressure fell below 12 bar (12 times modern sea-level pressure).

With life clearly, even dominantly existing on virtually every celestial object, the notion that life originated on Earth is absurd. The notion that the rest of the universe is dead is laughable. And the concept of error-based mutationary evolution is unnecessary.

The keys of evolution are in mankind's hands at this present time, and those keys are genetic engineering.

The Law of Gravity

The Law of Gravity is repaired with the trust angle of Earth and completed with the Law of Time.

The Law of Time, through recognizing the relatively minor mass of Sagittarius A*, recognizes that the Milky Way is not held together by Sag A*. That, plus resolving that the value of G is over three times greater than formerly thought, eliminates the need for dark matter as holding the Milky Way together. Dark matter does not exist. Dark matter was imagined in an attempt to finish or amend the Law of Gravity.

The mass of Sagittarius A* is so low that it has virtually no gravitational effect on any other galaxy – not even Andromeda. Since there is virtually no attraction between galaxies, there is no need for dark energy pushing them apart.

The information which was carefully cherry-picked to suggest that the universe is expanding or experiencing accelerating expansion is sheer misinterpretation of spectroscopic data and nothing more. The associated dark energy is also fictional. Dark energy was imagined in an attempt to finish or amend the Law of Gravity.

The Law of Gravity must be scaled using the Law of Time and a different value for gravity.

To avoid confusion of rewriting a popular standard, the new gravity constant is the James Newton, N, which equals 2.277E-10.

$$\text{N} = 2.277\text{E-}10$$

The New Age of Physics is **The Age of Paradox Null**.